Wilbur Olin Atwater

Methods and results of investigations on the chemistry and economy of food

Wilbur Olin Atwater

Methods and results of investigations on the chemistry and economy of food

ISBN/EAN: 9783337201210

Printed in Europe, USA, Canada, Australia, Japan

Cover: Foto ©berggeist007 / pixelio.de

More available books at **www.hansebooks.com**

U. S. DEPARTMENT OF AGRICULTURE.

OFFICE OF EXPERIMENT STATIONS.

METHODS AND RESULTS

OF

INVESTIGATIONS

ON THE

CHEMISTRY AND ECONOMY OF FOOD.

BY

W. O. ATWATER, Ph. D.,

Professor of Chemistry in Wesleyan University,
Director of the Storrs (Conn.) Agricultural Experiment Station, and
Special Agent of the United States Department of Agriculture.

PUBLISHED BY AUTHORITY OF THE SECRETARY OF AGRICULTURE.

WASHINGTON:
GOVERNMENT PRINTING OFFICE.
1895.

CONTENTS.

LIST OF ILLUSTRATIONS.

LETTER OF TRANSMITTAL.

UNITED STATES DEPARTMENT OF AGRICULTURE,
OFFICE OF EXPERIMENT STATIONS,
Washington, D. C., October 20, 1894.

SIR: I have the honor to transmit herewith for publication Bulletin No. 21 of this Office, on "Methods and Results of Investigations on the Chemistry and Economy of Food," by W. O. Atwater, Ph. D. The circumstances which have led to the publication of the present bulletin need, perhaps, some words of explanation. Investigations of the hygienic and pecuniary economy of food are of comparatively recent date. It is scarcely fifty years since the classical researches of Liebig began to pave the way for finding practically all we know to-day of the ingredients of our food materials, the ways in which they are used in the body, and the kinds and combinations which are best adapted to health and purse. Nearly all of the best experimental inquiry in these lines has been carried on in Europe. There have, indeed, been pioneers in the United States, but their work belongs mostly to the last two, and all of it to the last three, decades. The first analyses by modern methods of materials used for the food of man or domestic animals in this country were made in 1869 by the author of this bulletin, then a student in the Sheffield Scientific School of Yale University. With the rise of the experiment stations, inquiries into the composition of feeding stuffs and their appropriate use in the nutrition of domestic animals have been undertaken and during the past few years have been carried on quite actively. The first at all extensive series of investigations of materials used as the food of man undertaken in the United States were studies of the chemistry of fish, prosecuted under the auspices of the United States Fish Commission in the chemical laboratory of Wesleyan University by Professor Atwater in the years 1878–1881. In connection with this work, analyses of meats and other food materials were made under the auspices of the Smithsonian Institution.

The first accurate investigation of the chemical and economical statistics of food consumption in the United States were undertaken in the year 1886 by Col. Carroll D. Wright, then chief of the Massachusetts Bureau of Statistics of Labor and now United States Commissioner of Labor, in cooperation with Professor Atwater at Middletown, Conn. Thus far but extremely little has been done in the United States by way of exact experimental inquiry regarding the laws of human nutrition. But the growth of the preliminary forms of investigation points to a speedy development of research of a high character in this direction also. It is a very gratifying fact that already a considerable amount of investigation regarding the composition of our food materials and the economy of their use has accumulated, and, what is more to the point, earnest and capable investigators are addressing themselves to the still more extended and thorough study of the subject.

A large part of the work thus far done in the United States has been at private expense. But, as often happens, the inquiries thus benevolently begun have proven so useful that public funds are becoming available for their prosecution. On the recommendation of the Secretary of Agriculture the sum of $10,000 was included in the appropriations for the Department of Agriculture for the present fiscal year, the purpose of which is to enable him to investigate and report upon the food economy of the people of the United States. The supervision of the investigations thus provided for has been assigned to the Office of Experiment Stations, and Professor Atwater has been appointed special agent in charge.

The agricultural experiment stations established by authority of Congress are authorized to make inquiries in this same direction, and are called upon to report their results to the Secretary of Agriculture. In order to facilitate such investigations, not only by investigators in experiment stations, but by those associated with the colleges and other institutions for research and instruction, it has seemed desirable to prepare a statement of the results of inquiries already reached, questions which now demand attention, and the best methods for their study. It was, however, found impracticable to make a complete presentation of the subject without greatly delaying the publication of such information as seemed almost essential to the further prosecution of researches in this line in this country. An effort has therefore been made to outline in a general way the field to be traversed, to show with detail sufficient for present purposes what portions need first to be cultivated and to indicate how the work should be carried on. It is hoped that the remaining material may in due time be collated and published.

Respectfully,

A. C. TRUE, *Director.*

HON. CHAS. W. DABNEY, Jr.,
Acting Secretary of Agriculture.

THE CHEMISTRY AND ECONOMY OF FOOD.

CHAPTER I.

INTRODUCTION.

PUBLIC NEED OF INFORMATION ABOUT FOOD ECONOMY.

Food constitutes the chief item of the living expenses of the people, of our agricultural production, and of our exports. Half the earnings of wage workers in this country and in Europe is spent for food. The health and strength of all are intimately dependent upon their diet. Yet most people understand very little about what their food contains, how it nourishes them, whether they are economical or wasteful in buying and preparing it for use, and whether or not the food they eat is rightly fitted to the demands of their bodies. The result of this ignorance is great waste in the purchase and use of food, loss of money, and injury to health.

Underneath all this is still another evil. Our food products as a whole are not such as are best fitted for healthful and economic nutrition or for the most profitable export. Our agricultural production is out of balance. Here, again, the facts are not clearly understood, as they must be to make the needed reform possible.

The reason for this ignorance is simple enough. Fifty years ago no man knew what our bodies and our foods were composed of; how the different nutritive ingredients of the food served their purposes in nutrition; how much of each of the ingredients was needed to supply the demands of people of different age, sex, and occupation; and how best to adjust the diet to the wants of the user. We do not to-day know as much about these things as we ought. For that matter, we never shall be able to lay down hard and fast rules to apply to all cases, because of the differences between individuals in respect to their demands for nutriment and the ways in which their bodies can make use of different kinds of foods. But the research of the past twenty-five years has brought a great deal of definite information. Nearly all of the exact inquiry in this direction has been done in Europe, and the greater part of it in Germany. We are beginning it in the United States.

To remove this ignorance two things are needed. The first is a more definite knowledge of the actual facts. The second is that the information be brought to the people. The knowledge can be gained by research. To secure its diffusion the results of inquiry must be published in detail. They can then be popularized and made useful to the people at large.

PURPOSE AND PLAN OF THIS BULLETIN.

This bulletin has a twofold purpose: to summarize some of the results of late inquiry as to the physiological and pecuniary economy of food, and to indicate questions now demanding study and desirable methods of investigation.

Among the results of inquiry dwelt upon are the chemical composition of materials used for the food of man; the proportions of nutritive ingredients; their digestibility; their fuel values; the ratios between their values for nutriment and their cost; the kinds of food and proportions of nutrients best adapted to the demands of people of different classes and occupations; the errors in our food economy; and the sociological and agricultural bearings of the subject.

The questions proposed for study are such as the above-named topics suggest. They center around the general problem of food economy. Some are abstrusely scientific and reach into the higher realms of chemistry and physiology, but these are as necessary for the practical inquiry as the foundation is for the house. All bear directly or indirectly upon health, home life, and household, agricultural, and national economy. The methods of inquiry recommended are mainly such as have been elaborated more or less completely by experience and described succinctly by specialists. Some, however, are new and still in the process of development.

COLLABORATORS.

In the preparation of this bulletin the author has enjoyed the valuable assistance of Messrs. C. D. Woods, H. B. Gibson, and C. F. Langworthy. Mr. Woods has performed a large part of the labor of compiling the analyses of foods, and the results of his experience in the study of methods and the actual work of analysis in the laboratory during the last fifteen years or more are embodied in the description of methods of analysis recommended. He has also assisted in the studies of dietaries and compilation of results. Dr. Gibson, who has likewise shared in the studies of foods and dietaries for several years, has done a large part of the work of compiling the results of such studies. Dr. Langworthy has shared in the collating of studies of dietaries and has been especially helpful in the collating of the results of investigations of the digestibility of foods.

CHAPTER II.

FOOD AND ITS USES FOR NUTRIMENT.

A pound of lean beef and a quart of whole milk contain about the same amounts of actually nutritive material. But the pound of beef costs more than the quart of milk, and its nutrients differ not only in number and kind, but are, for ordinary use, more valuable than those of the milk. This illustrates a fundamental fact in the economy of foods, namely, that the differences in the values of different foods depend upon both the kinds and the amounts of the nutritive materials which they contain. Add to this that it is essential for health that the food shall supply the nutrients in the kinds and the proportions required by the body, and that it is likewise important, from a pecuniary standpoint, that the materials be obtained at the minimum cost, and we have the fundamental principles of food economy.

COMPOSITION OF FOOD MATERIALS—NUTRIENTS AND THEIR FUNCTIONS IN NUTRITION.

Ordinary food materials—such as meat, fish, eggs, potatoes, wheat, etc.—may be regarded as consisting of refuse and edible portion.

Refuse.—This includes the bones of meat and fish, shells of shellfish, skin of potatoes, bran of wheat, etc.

Edible portion.—This includes the flesh of meat and fish, the white and yolk of eggs, wheat flour, etc. The edible portion consists of *water* and nutritive ingredients, or *nutrients.*

Current usage recognizes the following as the principal kinds of nutrients: Protein, fats, carbohydrates, and mineral matters.

The water, refuse, and salt of salted meat and fish are here designated as "non-nutrients." The water is, of course, indispensable for nourishment, but is not a nutrient in the sense in which the word is here used. In comparing the values of different food materials for nourishment the refuse and water are left out of account.

CLASSES OF NUTRIENTS.

The following familiar examples of compounds commonly grouped with each of the four principal classes of nutrients will serve to define the terms as here used, and may perhaps help to avoid the confusion which unfortunately results from the variations in usage by different writers:

Protein	Proteids	Albuminoids: e. g., albumen of eggs; myosin, the basis of muscle (lean meat); the albuminoids which make up the gluten of wheat, etc.
		Gelatinoids: Constituents of connective tissue which yield gelatin and allied substances, e. g., collagen of tendon; ossein of bone.
	"Nitrogenous extractives" of flesh, i. e., of meats and fish. These include kreatin and allied compounds, and are the chief ingredients of beef tea and most meat extracts.	
	Amids: This term is frequently applied to the nitrogenous non-albuminoid compounds of vegetable foods and feeding stuffs, among which are amido acids, such as aspartic acid and asparagin. Some of them are more or less allied in chemical constitution to the nitrogenous extractives of flesh.	
Fats	Fat of meat; fat of milk; oil of corn, wheat, etc. The ingredients of the "ether extract" of animal and vegetable foods and feeding stuffs, which it is customary to group together roughly as fats, include, with the true fats, various other substances, as lecithins and chlorophylls.	
Carbohydrates	Sugars, starches, celluloses, gums, woody fiber, etc.	
Mineral matters	Potassium, sodium, calcium and magnesium chlorids, sulphates and phosphates.	

An illustration of the prevalent confusion in the use of terms is found in those for the nitrogenous compounds. The words "albuminoids" and "proteids" are employed by different authors with very different significations. Some, for instance, apply the former of these terms to what are here called "gelatinoids." Others make a somewhat similar use of the term "proteids." Still others use one or the other of these terms for the whole nitrogenous material, which is here designated as "protein."

USES OF FOOD—FUNCTIONS OF NUTRIENTS.

The two chief uses of food of animals are: First, to form the materials of the body and repair its wastes; and, second, to yield energy in the forms of (1) heat to keep the body warm and (2) muscular and other power for the work it has to do. In forming the tissues and fluids of the body the food serves for building and repair. In yielding energy it serves as fuel for yielding heat and power.

The different nutrients of food act in different ways in fulfilling these purposes. The principal tissue formers are the albuminoids. These form the framework of the body. They build and repair the nitrogenous materials, as those of muscle, tendon, and bone, and supply the albuminoids of blood, milk, and other fluids. The chief fuel ingredients of the food are the carbohydrates and fats. These are either consumed in the body or are stored as fat to be used as occasion demands.

BUILDING AND REPAIR—FUNCTIONS OF THE PROTEIN COMPOUNDS.

The albuminoids are the building material of the body. The bodily machine is made from them, but in the making of the machine the albuminoids remain partly albuminoids and are partly changed to gelatinoids, so that the machine, as built, consists of both albuminoids and gelatinoids. The gelatinoids can not, according to the best evidence now at hand, be transformed into albuminoids, but they do serve to protect the albuminoids from being consumed. Both albuminoids and gelatinoids, after they have served as building material, can be broken up and oxidized within the body. In this cleavage and oxidation they serve as fuel.

Still another function of the protein is the formation of fats and carbohydrates. These latter are produced by the cleavage of the molecules of the proteid compounds. It is reasonably certain that the albuminoids, and probable, or at any rate possible, that the gelatinoids also, are thus transformed in the animal organism. Similar processes appear to take place in non-living protein compounds (e. g., cheese) under the influence of ferments.

The nitrogenous extractives can neither build tissue nor serve as fuel, but they are useful otherwise. Just how they are useful is not yet fully explained, but they appear to exert some influence upon the nervous system, to act as stimulants, and thus to help the body to make use of other materials in its nourishment.

The amids do not appear to serve any purpose as building material in the animal body. Like the nitrogenous extractives, they are products of the cleavage of the more highly organized proteids. But while they do not appear to be used for either building or repair, they, or some of them at least, serve as fuel, and it is possible that they may, like the gelatinoids, help to protect the albuminoids of the food and of the body tissues from being consumed.

The albuminoids are the most important of the protein compounds, both because they are the only ones that are actually used for building material and because they make up the bulk of the protein of the food and of the body. Gelatinoids and nitrogenous extractives occur only in such animal tissues as muscle, tendon, etc., and their quantity in these is small. The amids are found in considerable quantities in tubers, as potatoes; in roots, as turnips and beets, and in fruits; they are not found to any extent in other food materials. Since the quantities of gelatinoids, nitrogenous extractives, and amids in our food materials are so small, we do not go far astray in following the ordinary practice of using the term protein to denote the building material of the food.[1]

FUEL—FUNCTIONS OF FATS AND CARBOHYDRATES.

The machine needs fuel. Starch and sugar are burned in the body and yield heat and power, just as truly as does the coal which is burned

[1] See also pp. 44–49.

in a stove to heat the house or under a boiler to drive an engine. The fats serve the same purpose, only they are more concentrated fuel than the carbohydrates. The body transforms the carbohydrates into fat, which it keeps as a reserve of fuel in the most concentrated form. While the fat of the food is consumed more or less directly, part of it is stored as fat in the body. At the same time the previously stored body fat is being drawn upon for use as fuel. The carbohydrates of the food are consumed more or less directly in the body. Small quantities are transformed into fat, as above stated, and other quantities, probably in most cases still smaller, appear to be transformed into the carbohydrates of the body. The quantity of carbohydrates in the body is at most quite small. The principal one is glycogen. Inosit, which was formerly reckoned with the carbohydrates, is found to have a different constitution, and to contain a benzene nucleus.

The fats and carbohydrates are not the only materials that can be used us fuel. The protein compounds can perform the same service. A dog can live on lean meat, which thus serves as both building material and fuel. We can likewise use the protein of our bodies to supply us with both heat and muscular strength. This last statement may be expressed in another form so as to emphasize an important difference between the protein and the other ingredients of food and between the animal machine and other machines. The protein compounds can do the work of the carbohydrates and fats in being consumed for fuel, but the carbohydrates and fats can not do the work of protein in building and repairing the tissues of the body.

The bodily machine is made of protein. That is to say, blood, muscle, tendon, bone, and brain, all consist of, or at least contain, protein compounds. These are formed from the myosin of meat and fish, the casein of milk, the albumen of eggs, the gluten of wheat, and other albuminoids of the food. As the muscles and other tissues are used up in bodily activity, the same materials of the food are used for their repair. Of course, the mineral matters have a good deal to do with the building up of the tissues. Thus, phosphate of lime is an essential ingredient of the bones.

The chief fuel materials of the bodily machine are carbohydrates and fats, but the protein of the food and the tissues also serves as fuel.

The animal machine differs from others in that it can use its own substance for fuel.

THE POTENTIAL ENERGY OF FOOD—FUEL VALUE.

Heat and muscular power, like mechanical power, light, and electricity, are forms of energy. The energy is latent in the food, and is developed as the food is consumed in the body. We call it potential energy, and measure its quantity as we measure quantities of heat or of mechanical power. In other words, the value of food for fuel is expressed in terms of potential energy. The quantities of potential energy in food materials are determined by the calorimeter. The unit commonly used is the calorie,[1] the amount of heat which would raise the temperature of a kilogram of water 1° C. (or 1 pound of water 4° F.). Instead of this unit of heat we may use a unit of mechanical energy, for instance, the foot-ton, which is the force that would lift 1 ton 1 foot. One calorie corresponds very nearly to to 1.53 foot-tons.

Taking ordinary food materials as they come, the following general estimate has been made for the average fuel value of one gram of each of the classes of nutrients:

Fuel values of nutrients of food.

	Calories.	Foot-tons.
In 1 gram of protein	4.1	6.3
In 1 gram of fats	9.3	14.2
In 1 gram of carbohydrates	4.1	6.3

These figures mean that when a gram of fat, be it the fat of food or body fat, is consumed in the body, it will, if its potential energy be transformed into heat, yield enough to warm $9\frac{3}{10}$ kilograms of water 1° C. (or 41 pounds of water 1° F.); or, if it be transformed into mechanical energy, such as the steam engine or the muscles use to do their work, it will furnish as much as would raise 1 ton $14\frac{1}{5}$ feet, or $14\frac{1}{5}$ tons 1 foot. A gram of protein or carbohydrates would yield a little less than half as much energy as a gram of fat.

It must be remembered that we are to-day only at the beginning of our knowledge of these subjects, and a very large amount of the most laborious, painstaking, and abstract inquiry will be needed to tell satisfactorily what needs to be known. But we are doubtless warranted in using these figures with the distinct understanding that they are tentative and subject to such revision as future research shall indicate. In other words, when we compare the nutrients in respect to their fuel values, their capacities for yielding heat and mechanical power, an ounce of protein of lean meat or albumen of egg is just about equivalent to an ounce of sugar or starch; and a little over 2 ounces of either would be required to equal an ounce of the fat of meat or butter or the body fat. The potential energy in the ounce of protein or carbohydrates would, if transformed into heat, suffice to raise the temperature of 113 pounds of water 1° F., while an ounce of fat, if completely burned in the body or in the calorimeter, would yield as much heat as would warm over twice that weight of water 1°.

[1] The same word "calorie" is used to designate the heat required to raise the temperature of a gram of water a degree. The large Calorie is commonly spelled with a capital "C." One Calorie is thus equal to 1,000 calories. It does not seem necessary, however, to maintain this distinction at all times, and the spelling "calorie" will be used for the large Calorie in this publication.

The figures for fuel value in the tables of composition of food materials beyond are calculated for each material by multiplying the number of grams of protein and of carbohydrates in 1 pound (1 pound equals 453.6 grams) by 4.1 and the number of grams of fat by 9.3, and taking the sum of these three products as representing the number of calories in a pound of the material. The figures for fuel value in the dietaries beyond are computed in like manner.

The methods and results of experimental inquiry regarding this general subject will be referred to again.

The functions of food and its nutrients may be briefly summarized as follows:

WAYS IN WHICH FOOD IS USED IN THE BODY.

Food supplies the wants of the body in several ways. It either (1) is used to form the tissues and fluids of the body; (2) is used to repair the wastes of tissues; (3) is stored in the body for future consumption; (4) is consumed as fuel, its potential energy being transformed into heat or muscular energy or other forms of energy required by the body; or, (5) in being consumed protects tissues or other food from consumption.

USES OF THE DIFFERENT CLASSES OF NUTRIENTS.

<table>
<tr><td>Protein forms tissue (muscle, tendon, etc., and fat) and serves as fuel.</td><td rowspan="3">All yield energy in form of heat and muscular strength.</td></tr>
<tr><td>Fats form fatty tissue (not muscle, etc.) and serve as fuel.</td></tr>
<tr><td>Carbohydrates are transformed into fat and serve as fuel.</td></tr>
</table>

In being themselves burned to yield energy, the nutrients protect each other from being consumed. The protein and fats of body tissue are used like those of food. An important use of the carbohydrates and fats is to protect protein (muscle, etc.) from consumption.

DEFINITION OF FOOD.

In this view food may be defined as material which, when taken into the body, serves to either form tissue or yield energy, or both. This definition includes all the ordinary food materials, since they both build tissue and yield energy. It includes sugar and starch, because they yield energy and form fatty tissue. It includes alcohol, because the latter is burned to yield energy, though it does not build tissue. It excludes creatin, creatinin, and other so-called nitrogenous extractives of meat, and likewise thein or caffein of tea and coffee, because they neither build tissue nor yield energy, although they may, at times, be useful aids to nutrition.

CHAPTER III.

COMPOSITION OF FOOD MATERIALS.

Nearly all of our definite knowledge of the chemical composition of food materials and their nutritive values has accumulated within comparatively few years. Yet so active has been the inquiry, especially in the last one or two decades, that the data are to-day quite extensive.

HISTORICAL DEVELOPMENT OF FOOD ANALYSIS.

The first effective impulse to the systematic investigation of the chemistry of food was given by Liebig. It was he who distinguished clearly between the different constituents and indicated their several uses in nutrition. He attributed relatively more importance to the nitrogenous ingredients than later research has justified. He taught that they are the chief sources of muscular force, a doctrine which few would accept to-day. But in the main his classification of the nutrients and their functions has stood the severe test of forty years of experiment and criticism. In the hands of his followers the methods of analysis have been developed and more accurate distinctions have been made between different classes of ingredients. The individual compounds and the changes they undergo in the plant and in the animal have been studied and, with the rapid spread of research in organic, physiological, and agricultural chemistry, in physiology, and in bacteriology, new fields of inquiry have been opened up and research is being pushed forward with great activity.

An adequate historical view of the subject would hardly be in place here. The progress may be clearly seen by referring to the works on food and nutrition which have been the standards at various epochs, beginning with the early editions of Liebig's works on physiological and agricultural chemistry which appeared in 1842 and succeeding years, and including Dietrich and König's magnificent compilations of the results of investigations upon composition and digestibility of feeding stuffs for domestic animals[1] and König's equally valuable compilation regarding the foods and food adjuncts used by man.[2]

The compilations just referred to illustrate most strikingly the accumulation of data in Europe during comparatively few years. The first edition of Dietrich and König's compilation of analyses of feeding stuffs

[1] Dietrich and König. Zusammensetzung und Verdaulichkeit der Futtermittel. Zweite Auflage; Bde. I and II; Berlin, 1891.

[2] J. König. Chemie der menschlichen Nahrungs- und Genussmittel. Dritte Auflage; Bde. I and II; Berlin, 1889–1893.

was published in 1874. It is a thin volume of between 90 and 100 pages, consisting mostly of tabular statements of results of analyses with references to the original publications, and very brief comments. The second edition, seventeen years after, likewise in quarto form, is in two volumes. These contain 1,433 pages. Nearly the whole is devoted to statements, mostly in figures, of results of analyses and of experiments upon digestibility of feeding stuffs.

König's compilation regarding the chemistry of materials used as the food of man has lately reached its third edition. The first volume contains 1,189 pages. Introductory statements fill 184 pages of this volume; the rest is devoted to tabular statements of results of analyses and brief comments. The second volume includes 1,401 pages, and is devoted mainly to detailed descriptions of the different kinds of food materials.

While it is hardly possible to exaggerate the value of these monuments of German painstaking scholarship, they are very far from meeting the needs of English-speaking people, especially in the United States, not only because they are in the German language and written for German readers, but also because much that people in this country who are interested in such matters want to know is not contained in them. Our food materials, our habits of food consumption, and our food production all differ to a greater or less extent from those in Germany, and we need information fitted to our special circumstances.

It would be hardly proper to omit from even so brief an account of the development of food analysis as this a reference to what has been done since the introduction of the so-called Weende method, of which the late Professor Henneberg was the author. This dates practically from the year 1864, at which time a plan for methods for analysis of feeding stuffs was presented at the second Convention of German Agricultural Chemists.[1]

As long ago as 1836 Boussingault reported analyses[2] of a number of feeding stuffs, laying especial stress upon the quantities of nitrogen. In 1831 Boussingault and Le Bel reported analyses of milk.[3] The methods, however, were quite imperfect. For a number of years the chief stress was laid upon the quantities of carbon and nitrogen. Even in the works of Payen, as late as 1865, the simpler methods which took into account little else than the water, nitrogen, and carbon were followed, although more or less effort was made to determine the quanti-

[1] Landw. Vers. Stat., 6, 496.

[2] Recherches sur la Quantité d'Azote contenue dans les Fourrages, et sur leurs Équivalens; par M. Boussingault. Ann. Chim. et Phys., vol. 63, p. 225, 1836. Recherches sur la Quantité d'Azote contenue dans les Fourrages, et sur leurs Équivalens; par M. Boussingault. Ann. Chim. et Phys., vol. 67, p. 408, 1838. Ibid., 68, p. 408.

[3] Recherches sur l'Influence de la Nourriture des Vaches, sur la Quantité et la Constitution chimique du Lait; par MM. Boussingault et Le Bel. Ann. Chim. et Phys., vol. 71, p. 65, 1839.

ties of fats and to distinguish between the different nitrogenous compounds, as well as the carbohydrates.[1]

A distinct advance is found in some of the publications between 1840 and 1864, but it was not until the Weende method came into use and compilations of results of analyses made in accordance with it had become current that chemists generally adopted the forms which are now in general use.

ANALYSES OF AMERICAN FOOD PRODUCTS.

The first analyses made by these modern methods in the United States were a series of analyses of Indian corn by the writer under the direction of Prof. S. W. Johnson in New Haven in 1869.[2]

With the exception of the excellent work by Professor Storer at the Bussey Institution, comparatively little was done in this direction until the establishment of the experiment stations. The station at Middletown, Conn., made a number of analyses in the years 1877–1879. Others followed in rapid succession, though nearly all of the attention was given to feeding stuffs. The compilation of analyses of American feeding stuffs by Jenkins and Winton,[3] which "was designed to include all analyses published before September 1, 1890," and does include very nearly all, contains results of examinations of 3,267 specimens. The condensed tabular compilation of these fills 138 pages.

The compilation just referred to contains very few analyses of materials used as the food of man, and it is not until within a short time past that much attention has been given to this particular branch of investigation. A large number of analyses, indeed, have been made of milk, butter, and cheese, but the purpose of these has been mainly agricultural. A considerable number of analyses of fruits and vegetables have also been made by the Bussey Institution, the Division of Chemistry of the United States Department of Agriculture, and the Maine, Massachusetts, Connecticut, New York, Tennessee, California, and other experiment stations. The first at all extended series of examinations of materials used as food from the standpoint of their nutritive value for man in the United States was a study of American fishes and the invertebrates. This was undertaken at the instance and partly at the expense of the Smithsonian Institution and the United States Fish Commission through the influence of the late Prof. S. F. Baird. The work was performed in the chemical laboratory of Wesleyan University under the direction of the writer. In connection with this, at the instance of the United States National Museum, a series of analyses of American meats was undertaken. These investigations were carried out during the years 1878–1891, and included analyses of

[1] See, for instance, Substances Alimentaires. Payen, 1865.

[2] Am. Jour. Sci., vol. 48, 1869, p. 352.

[3] A Compilation of Analyses of American Feeding Stuffs, by E. H. Jenkins, Ph. D., and E. L. Winton, Ph. B. U. S. Dept. Agr., Office of Experiment Stations, Bul. No. 11. Washington, 1892.

the flesh of some 200 specimens of sea and fresh-water fishes, oysters, clams, etc., and of nearly 100 specimens of meats. A few analyses of milk, butter, cheese, flour, bread, etc., were also included. The results of the analyses of fishes and the invertebrates have been published by the United States Commission of Fish and Fisheries,[1] those of meats and vegetable foods were published by the Storrs (Conn.) Experiment Station.[2]

The number of analyses of dairy products in the United States has grown to be very large—so large indeed that complete compilation would be extremely difficult and perhaps of doubtful value.

A considerable number of specimens of materials used for the food of man, especially of canned vegetables, have been made in the Division of Chemistry of the Department of Agriculture under the direction of Prof. H. W. Wiley.[3] Quite a number of analyses have also been made in connection with the studies of adulteration of foods in the same laboratory. Some years previous a considerable number of analyses of specimens of American flour and the bread made from them were made in the same laboratory by Mr. Clifford Richardson.[4]

A series of analyses of milling products of wheat was made by Professor Kedzie in Michigan some years ago.[5]

In his report on cereals, which was published in the volume on agriculture of the Report of the United States Census for 1880, Prof. W. H. Brewer has given analyses of a number of cereal products. During the years 1890–1894, in connection with studies of dietaries by the Storrs (Conn.) Experiment Station, a number of analyses of food materials have been made. These, however, have been published only in part.

The collection of food materials at the World's Fair at Chicago, in 1893, offered an unusually favorable opportunity for obtaining specimens for chemical examination. This fact was appreciated by the World's Columbian Commission, and Prof. H. W. Wiley, of the United States Department of Agriculture, was called upon by the Bureau of Awards of that Commission to execute analyses of a large number of specimens of grains and milling products from them. These investigations are being continued in the chemical laboratory of that Department, and are not fully completed at the time of this writing.

At the same time the writer was invited by the Bureau of Awards to make examinations of such other food materials as seemed advisable. The remarkably favorable occasion for securing specimens of meats and meat products as prepared at the great slaughtering establishments in Chicago was also utilized. Some 500 such specimens of food materials

---

[1] The Chemical Composition and Nutritive Values of Food-Fishes and Aquatic Invertebrates, by W. O. Atwater, Ph. D. Report U. S. Commissioner of Fish and Fisheries, 1888. Washington, 1891.

[2] W. O. Atwater and C. D. Woods. Reports Storrs (Conn.) Experiment Station, 1891, 1892, and 1893, passim.

[3] Bull. 13, Div. Chem., U. S. Dept. Agr., pp. 1015–1167.

[4] Bull. 4, Div. Chem., U. S. Dept. Agr.

[5] Rept. Mich. Bd. Agr., 1877, p. 350.

from the exhibits by the various nations at the fair in Chicago and from the slaughtering establishments in the city were collected and analyzed. Part of the work was done in Chicago during the fair, and the rest was completed later in the chemical laboratory of Wesleyan University, in connection with the work of the Storrs Station, which is conducted there. The results of these analyses still await publication.

TABLES OF COMPOSITION OF AMERICAN FOOD MATERIALS.

The compilation of the results of analyses of American food materials up to the present time is greatly to be desired. An effort in this direction is already begun, but the results are not yet ready for publication. Tables 1 and 2, which follow, give a summary of the results thus far published. It is essentially the same as that published in the report of the Storrs (Conn.) Experiment Station for 1891.[1] To the figures there given have been added the results of examinations of specimens of canned vegetables by Professor Wiley and his assistants which were previously mentioned.

It will be observed that the table gives in the first column the number of analyses of specimens of each kind of material. In a large number of cases only a single analysis had been made at the time, or at any rate had been found by us at the time when the table was prepared. In other cases, several analyses are available. Where more than one analysis has been made, the maximum, minimum, and averages are given. It should be explained, however, that the figures for maximum and minimum indicate the largest or smallest percentage of each ingredient found in any case, so that the figures for maximum or minimum of a given material do not all refer to the same specimen.

The figures for fuel value give the estimated calories in the protein, fats, and carbohydrates in 1 pound of the material. These figures are not the result of direct determinations, but are found by multiplying the number of grams of protein, fats, and carbohydrates by the factors 4.1, 9.3, and 4.1, respectively, and adding the products as explained above. (See p. 15.) Table 1 gives the composition of the food materials as they are ordinarily purchased in the markets, including both the refuse and the edible portion, while Table 2 gives the composition of the edible portion after the removal of the refuse. In some animal foods, as milk, cheese, and oysters (shell contents), and most of the vegetable foods (such as flour and bread) there is no refuse, and consequently the analyses of such materials appear, for the most part, in Table 2 alone. The relative composition of a small number of common foods is graphically shown in Chart 1.

ANALYSES OF SIDES OF BEEF AND MUTTON.

People in general have very little idea how wide may be the differences in composition of different specimens of meat of the same kind, as, for instance, beef from different animals and different cuts of beef

[1] The Composition of Food Materials. By W. O. Atwater and Chas. D. Woods.

CHART 1.—COMPOSITION OF FOOD MATERIALS.

Nutritive ingredients, refuse, and fuel value.

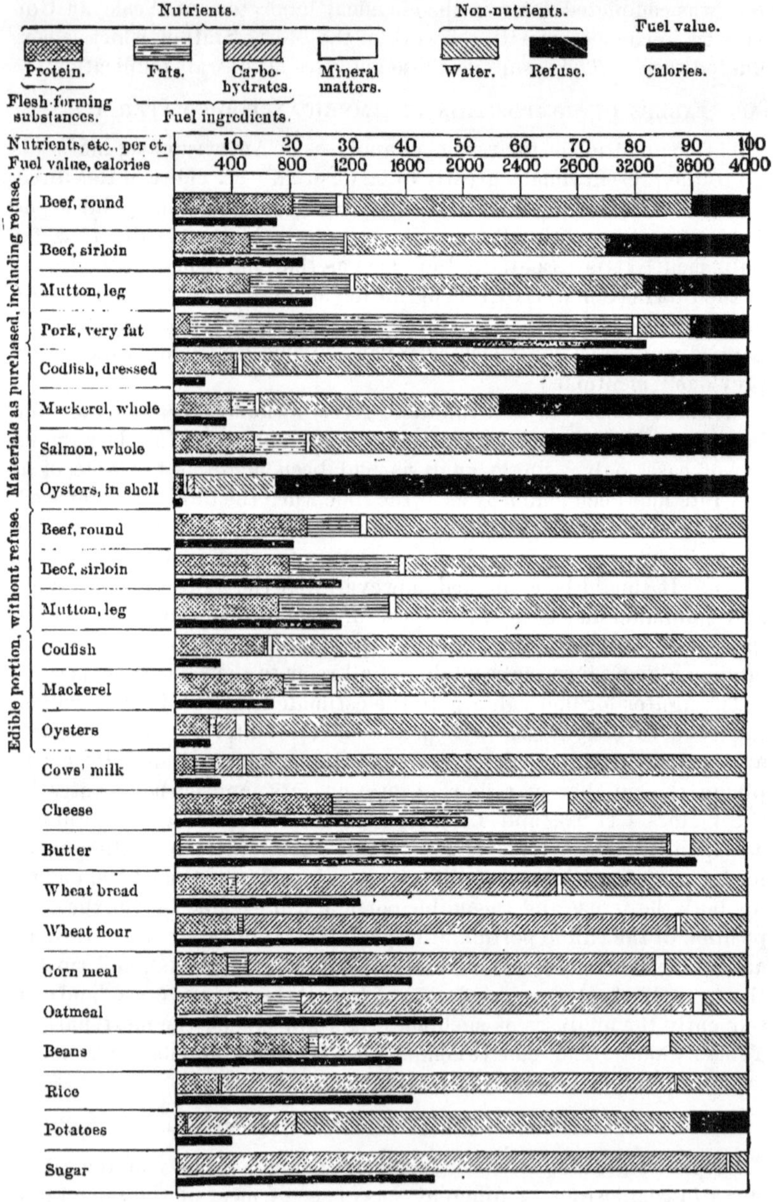

Nutrients. Non-nutrients. Fuel value.

Protein. Fats. Carbo-hydrates. Mineral matters. Water. Refuse. Calories.

Flesh-forming substances. Fuel ingredients.

Nutrients, etc., per ct. 10 20 30 40 50 60 70 80 90 100

Fuel value, calories 400 800 1200 1600 2000 2400 2600 3200 3600 4000

Materials as purchased, including refuse.

Beef, round
Beef, sirloin
Mutton, leg
Pork, very fat
Codfish, dressed
Mackerel, whole
Salmon, whole
Oysters, in shell

Edible portion, without refuse.

Beef, round
Beef, sirloin
Mutton, leg
Codfish
Mackerel
Oysters

Cows' milk
Cheese
Butter
Wheat bread
Wheat flour
Corn meal
Oatmeal
Beans
Rice
Potatoes
Sugar

from the same animal. These differences are illustrated in the analyses in Tables 1 and 2 below, and still more forcibly in Tables 3 to 10[1] and in fig. 1.

COMPOSITION OF SIDE OF BEEF.

Tables 3 to 6 give results of analyses of a side of beef which was butchered in Chicago and sold in Middletown, Conn. The side was called by the dealer in Middletown an average one as regards weight and fatness, etc. Its total weight at the car in the station at Middletown, where it was received, was 326 pounds, the fore quarter weight 176 pounds, and the hind quarter 150 pounds. It will be noticed in Table 6 that the sums of the weights of the individual cuts make the side weigh 317.6 pounds, and that the falling off in weight was almost wholly in the hind quarter, which weighed 150 pounds at the car and

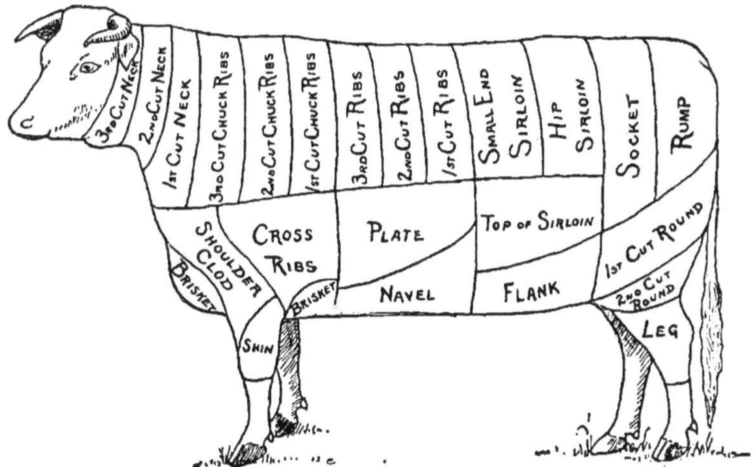

FIG. 1.—Diagram of cuts of beef.

by individual cuts only 142 pounds. The side, when received, was partly frozen, and the fore quarter was analyzed first. These two facts may account for the shrinkage in the weight of the hind quarter, which was kept for a longer time before the analyses were made. The cuts were made in accordance with the usage in New York markets, and are shown in the diagram herewith.

The fore quarter was thus divided into 15 and the hind quarter into 9 cuts. Each cut was weighed and a portion taken for analysis. The edible portion (flesh) and the refuse (bone, gristle, etc.) were separated and each weighed by itself. Table 3 shows the composition of the water-free substance of the edible portion: Table 4 the composition of the flesh (edible portion), and Table 5 the composition of the several cuts and of the whole side, including both refuse and edible portion.

[1] Taken from Report Storrs (Conn.) Experiment Station above referred to, pp. 59-65.

In Table 6 the results of the analyses of the whole side and the several cuts, which in Table 5 are represented in percentages, are calculated into pounds, so as to give the actual weights of the several ingredients in the different cuts and in the whole.

COMPOSITION OF SIDE OF MUTTON AND SIDE OF LAMB.

The side of mutton was from a Connecticut-grown sheep, 2 years old, of about medium fatness. One whole side was taken for analysis, the different cuts being made in the usual manner as specified in the tables.

The side of lamb was from a Connecticut-grown lamb 6 months old. The dressed weight was about 39 pounds. It was above rather than below medium fatness. The plan of the tables is the same as those for the side of beef.

TABLE 1.—*Composition of food materials as found in the markets, including both edible portion and refuse.*

Food materials	Number of analyses.	Salt.	Refuse (bones, skin, shell, etc.).	Edible portion.						
				Water.	Nutrients.					Fuel value of 1 pound.
					Total.	Protein.	Fat.	Carbohydrates.	Mineral matters.	
MEATS, ETC.		*P. ct.*	*P. ct.*	*P. ct.*	*P. ct.*	*P. ct.*	*P. ct.*	*P. ct.*	*P. ct.*	*Calories.*
Beef:										
Neck: Min	3	......	18.4	47	28.9	14.9	11.8	......	0.8	800
Neck: Max	3	......	22.9	52.4	32.1	16.2	16.4	......	.9	970
Neck: Avg	3	......	20	49.6	30.4	15.6	14	......	.8	880
Chuck ribs: Min	8	......	5.4	42.7	30.4	13.5	13.3	......	.6	865
Chuck ribs: Max	8	......	19.7	50.7	44.6	16.4	28.6	......	.9	1,490
Chuck ribs: Avg	8	......	14.6	49.5	35.9	15	20.1	......	.8	1,125
Ribs: Min	4	......	19.4	36.3	39.7	11.3	26.5	......	.7	1,360
Ribs: Max	4	......	22.7	40.2	42.3	13	29.2	......	.9	1,460
Ribs: Avg	4	......	21	38.2	40.8	12.2	27.9	......	.7	1,405
Brisket	1	......	14.3	40.6	45.1	12.5	31.9	......	.7	1,580
Cross ribs	1	......	12.2	38.6	49.2	12	36.5	......	.7	1,765
Shoulder: Min	6	......	10.7	49.9	27	14.8	8.8	......	.7	695
Shoulder: Max	6	......	15.6	60.6	38.8	18.6	19.2	......	1	1,155
Shoulder: Avg	6	......	12.6	55.8	31.6	17	15.7	......	.9	895
Shin	1	......	40.1	44.2	15.7	13.6	1.4	......	.7	310
Plate	1	......	17.9	36.4	45.7	12.6	32.4	......	.7	1,600
Navel	1	......	11.4	42.2	46.4	13.4	32.3	......	.7	1,610
Sirloin: Min	4	......	13	42.8	29.7	12.5	14	......	.7	920
Sirloin: Max	4	......	27.3	52.6	35	19.1	18.7	......	1.2	1,075
Sirloin: Avg	4	......	19.5	48.3	32.2	15	16.4	......	.8	970
Socket	1	......	35.8	36.7	27.5	10.7	16.1	......	.7	880
Rump: Min	2	......	6.5	33.7	40.9	12.3	25.1	......	.6	1,340
Rump: Max	2	......	16.2	52.6	50.1	15	37.2	......	.8	1,800
Rump: Avg	2	......	11.3	43.2	45.5	13.7	31.1	......	.7	1,570
Round, first cut	1	......	7.8	60.9	31.3	18	12.3	......	1	855
Round, second cut	1	......	32.1	47.2	20.7	14	5.8	......	.9	505
Leg	1	......	62.2	27.3	10.5	7.9	2.1	......	.5	235
Top of sirloin	1	......	3.2	40.9	55.9	12.9	42.3	......	.7	2.025
Flank	1	......	11.5	24.8	64.2	10.6	53	......	.6	2,435
Fore quarter	......	......	18.5	44.1	37.4	14.1	22.5	......	.8	1,210
Hind quarter	......	......	20.2	44.4	35.4	13.6	21	......	.8	1,140
Side without kidney fat	......	......	19.2	44.3	36.5	13.9	21.8	......	.8	1,180
Veal:										
Shoulder	1	......	11.5	63.7	24.8	18.3	5.5	......	1	570
Shoulder and flank	1	......	24.3	49.7	26	14.9	10.2	......	.9	715
Mutton:										
Shoulder	1	......	16.3	49	34.7	15.1	18.8	......	.8	1,075
Breast	1	......	14.9	32	53.1	12.1	40.1	......	.9	1,915
Neck	1	......	27.6	40.3	32.1	11.7	19.8	......	.6	1,055
Rack	1	......	19.3	44.3	36.4	14.9	20.9	......	.6	1,160
Leg	1	......	18.1	50.6	31.3	15	15.6	......	.7	935
Loin	1	......	15.8	41.5	42.7	12.6	29.5	......	.6	1,480
Flank	1	......	2.1	37.9	60	15.4	44.1	......	.5	2,145
Fore quarter	......	......	19	42.3	38.7	13.7	24.2	......	.8	1,275
Hind quarter	......	......	15.7	46.1	38.2	14.3	23.2	......	.7	1,245
Side without kidney fat	......	......	17.3	44.2	38.5	14	23.7	......	.8	1.260

TABLE 1.—*Composition of food materials as found in the markets, etc.*—Continued.

Food materials.	Number of analyses.	Salt.	Refuse (bones, skin, shell, etc.).	Edible portion.						
				Water.	Nutrients.					Fuel value of 1 pound.
					Total.	Protein.	Fat.	Carbohydrates.	Mineral matters.	
MEATS, ETC.—continued.										
Lamb:		*P. ct.*	*P. ct.*	*P. ct.*	*P. ct.*	*P. ct.*	*P. ct.*	*P. ct.*	*P. ct.*	*Calories.*
Shoulder	1	......	20.3	41.3	38.4	14	23.6	....	.8	1,255
Breast	1	......	19.1	45.5	35.4	15.5	19.1	....	.8	1,095
Neck	1	......	17.7	46.7	35.6	14.4	20.4	....	.8	1,130
Leg	1	......	17.7	53.3	29	15.5	12.6	....	.9	820
Loin	1	......	12.2	48.1	39.7	16.7	22.1	....	.9	1,245
Fore quarter	......	......	18.8	44.7	36.5	14.7	21	....	.8	1,160
Hind quarter	......	......	15.7	51.3	33	16	16.1	....	.9	975
Side without kidney fat	......	......	17.8	47.9	34.8	15.3	18.6	....	.9	1,070
Pork:										
Shoulder roast — Min	4	......	7.1	38.8	39.6	13.2	25.4	....	.7	1,325
Shoulder roast — Max	4	......	20.3	49.4	45.9	14.1	28.9	....	.8	1,590
Shoulder roast — Avg	4	......	14.6	43	42.4	13.6	28	....	.8	1,435
Poultry, etc.:										
Chicken	1	......	38.2	41.6	17.2	15.1	1.2	....	.9	330
Turkey	1	......	32.4	44.7	22.9	16.1	5.9	....	.9	550
Hens' eggs in shell — Min	2	......	13.4	62.3	22.7	12.1	9.7	....	.8	635
Hens' eggs in shell — Max	2	......	13.9	63.9	23.8	12.2	10.8	....	.9	680
Hens' eggs in shell — Avg	2	......	13.7	63.1	23.2	12.1	10.2	....	.9	655
Preserved meats:										
Corned beef, rump	1	......	5	70.8	24.2	16.7	5.1	....	2.4	525
Corned beef, flank — Min	2	......	9.6	39	37.1	11.7	21.2	....	2.5	1,140
Corned beef, flank — Max	2	......	14.6	48.3	51.4	13.2	37.2	....	2.7	1,785
Corned beef, flank — Avg	2	......	12.1	43.7	44.2	12.4	29.2	....	2.6	1,460
Ham, salted and smoked	1	......	11.4	36.8	51.8	14.8	34.6	....	2.4	1,735
FISH, SHELLFISH, ETC.										
Fresh fish:										
Sturgeon, section	1	......	14.4	67.4	18.2	15.4	1.6	....	1.2	355
Red-horse, entrails removed	1	......	52.5	37.3	10.2	8.5	1.1	....	.6	205
Herring, whole	1	......	46	37.3	16.7	10	5.9	....	.8	435
Alewife, whole — Min	2	......	49.4	36.9	12.2	9.5	1.9	....	.8	255
Alewife, whole — Max	2	......	49.5	38.3	13.7	9.9	3	....	.8	310
Alewife, whole — Avg	2	......	49.5	37.5	13	9.7	2.5	....	.8	285
Shad, whole — Min	7	......	44.4	30.3	10.9	7.4	2.9	....	.6	260
Shad, whole — Max	7	......	58.8	39.5	18.6	10.5	7.3	....	.8	505
Shad, whole — Avg	7	......	50.1	35.2	14.7	9.2	4.8	....	.7	375
Smelt, whole — Min	2	......	34.8	39.0	11.1	9.6	.8	....	.7	215
Smelt, whole — Max	2	......	49	52.3	12.9	10.4	1.2	....	1.3	245
Smelt, whole — Avg	2	......	41.9	46.1	12	10	1	....	1	230
Whitefish, whole	1	......	53.5	32.5	14	10.3	3	....	.7	320
Ciscoe, whole	1	......	42.7	43.6	13.7	11	2	....	.7	290
California salmon, section — Min	2	......	....	57.9	31.8	16.1	14.8	....	.9	925
California salmon, section — Max	2	......	10.3	62.7	37.3	17	19.2	....	1.1	1,125
California salmon, section — Avg	2	......	5.2	60.3	34.5	16.5	17	....	1	1,025
Salmon, whole — Min	4	......	30.8	36.9	23.6	13.3	7.9	....	.9	610
Salmon, whole — Max	4	......	39.5	45	24.3	15.2	10	....	1	670
Salmon, whole — Avg	4	......	35.3	40.6	24.1	14.3	8.8	....	1	635
Salmon, entrails removed	1	......	23.8	51.2	25	14.6	9.5	....	.9	675
Lake trout, whole	1	......	56.3	30	13.7	7.7	5.4	....	.6	370
Lake trout, entrails removed	1	......	35.2	45	19.8	12.4	6.6	....	.8	510
Brook trout, whole — Min	3	......	45.2	38.6	11	9.2	.4	....	.5	225
Brook trout, whole — Max	3	......	59.1	43.8	12.3	10.2	1.5	....	.7	255
Brook trout, whole — Avg	3	......	48.1	40.4	11.5	9.8	1.1	....	.6	230
Pickerel, whole — Min	2	......	45.4	40.8	10.5	9.7	.2	....	.6	190
Pickerel, whole — Max	2	......	48.7	43.6	11	10	.3	....	.7	200
Pickerel, whole — Avg	2	......	47.1	42.2	10.7	9.8	.2	....	.7	195
Pike, whole	1	......	42.7	45.7	11.6	10.7	.3	....	.6	210
Muscalonge, whole	1	......	49.2	38.7	12.1	10	1.3	....	.8	240
Eel, salt water, dressed — Min	2	......	19	54.9	21.6	14.3	6.4	....	.7	535
Eel, salt water, dressed — Max	2	......	21.4	59.4	23.7	14.9	8.1	....	.9	620
Eel, salt water, dressed — Avg	2	......	20.2	57.2	22.6	14.6	7.2	....	.8	575
Mullet, whole	1	......	57.9	31.5	10.6	8.1	2	....	.5	235
Mackerel, whole — Min	5	......	33.8	35.8	12.2	8.4	1.4	....	.6	300
Mackerel, whole — Max	5	......	57.9	48.5	23.8	12.1	10.7	....	1	675
Mackerel, whole — Avg	5	......	44.6	40.4	15	10	4.3	....	.7	370
Mackerel, entrails removed	1	......	40.7	43.7	15.6	11.4	3.5	....	.7	360
Spanish mackerel, whole	1	......	34.6	44.5	20.9	13.7	6.2	....	1	515
Pompano, whole — Min	2	......	42.4	38.8	11.2	9.9	.8	....	.5	220
Pompano, whole — Max	2	......	48.6	40.2	18.8	10.5	7.8	....	.5	325
Pompano, whole — Avg	2	......	45.5	39.5	15	10.2	4.3	....	.5	370

TABLE 1.—*Composition of food materials as found in the markets, etc.*—Continued.

Food materials.	Number of analyses.	Salt.	Refuse (bones, skin, shell, etc.).	Edible portion.						
				Water.	Nutrients.					Fuel value of 1 pound.
					Total.	Protein.	Fat.	Carbohydrates.	Mineral matters.	
FISH, SHELLFISH, ETC.—cont'd.										
Fresh fish—Continued.		*P. ct.*	*P. ct.*	*P. ct.*	*P. ct.*	*P. ct.*	*P. ct.*	*P. ct.*	*P. ct.*	*Calories.*
Bluefish, entrails removed	1	......	48.6	40.3	11.1	9.8	.6	......	.7	205
Butter-fish, whole	1	......	42.8	40.1	17.1	10.2	6.3	......	.6	455
Black bass, whole. Min	2	......	53.6	34.6	9.4	8.5	.4	......	.5	175
Black bass, whole. Max	2	......	56	34.7	11.7	10	1.1	......	.6	230
Black bass, whole. Avg	2	......	54.8	34.6	10.6	9.3	.8	......	.5	205
Yellow perch, whole	1	......	62.7	30	7.3	6.7	.2	......	.4	135
Yellow perch, dressed	1	......	35.1	50.7	14.2	12.6	.7	......	.9	265
Wall-eyed pike, whole	1	......	57.2	34.1	8.7	7.9	.2	......	.6	155
Gray pike, whole	1	......	63.2	29.7	7.1	6.4	.3	......	.4	130
Striped bass, whole. Min	5	......	48.6	32.5	8.7	7.2	.7	......	.5	170
Striped bass, whole. Max	5	......	57.1	39.7	11.7	9.7	1.6	......	.6	240
Striped bass, whole. Avg	5	......	54.9	35.1	10	8.3	1.1	......	.6	200
Striped bass, entrails removed	1	......	51.2	37.4	11.4	8.7	2.2	......	.5	255
White perch, whole. Min	2	......	61.8	27.8	9	6.5	1	......	.4	185
White perch, whole. Max	2	......	63.2	28.9	9.3	7.8	2.1	......	.5	210
White perch, whole. Avg	2	......	62.5	28.4	9.1	7.2	1.5	......	.4	195
Sea bass, whole	1	......	56.1	34.8	9.1	8.3	.2	......	.6	160
Red grouper, entrails removed. Min	2	......	55.8	34.8	8.9	8.2	.2	......	.5	160
Red grouper, entrails removed. Max	2	......	55.9	35.8	9.3	8.5	.3	......	.5	170
Red grouper, entrails removed. Avg	2	......	55.9	35	9.1	8.4	.2	......	.5	165
Red snapper, whole	1	......	40	46.9	13.1	12	.4	......	.7	240
Red snapper, entrails removed. Min	2	......	45.3	36.8	10.7	9.2	.3	......	.6	200
Red snapper, entrails removed. Max	2	......	52.5	43.7	11	10	.9	......	.7	210
Red snapper, entrails removed. Avg	2	......	48.9	40.3	10.8	9.6	.6	......	.6	205
Porgy, whole. Min	3	......	57.3	27.8	7.1	6.1	.5	......	.5	135
Porgy, whole. Max	3	......	65.1	31.1	12	8.2	3.4	......	.6	295
Porgy, whole. Avg	3	......	60	29.9	10.1	7.4	2.1	......	.6	225
Sheepshead, whole	1	......	66	26.9	7.1	6.4	.2	......	.5	125
Sheepshead, entrails removed	1	......	56.5	31.3	12.2	8.8	2.9	......	.5	285
Red bass, whole	1	......	63.5	29.8	6.7	6.1	.2	......	.4	120
Kingfish, whole	1	......	56.6	34.4	9	8.1	.4	......	.5	170
Weakfish, whole	1	......	51.9	38	10.1	8.4	1.1	......	.6	200
Blackfish, whole. Min	2	......	56.2	29.2	6.7	6.3	.2	......	.2	125
Blackfish, whole. Max	2	......	64.1	33.7	10.1	8.3	1.2	......	.6	205
Blackfish, whole. Avg	2	......	60.1	31.5	8.4	7.3	.7	......	.4	165
Blackfish, entrails removed. Min	2	......	53.6	33.5	8.7	7.9	.4	......	.4	165
Blackfish, entrails removed. Max	2	......	57.8	36.4	10	8.7	.7	......	.6	190
Blackfish, entrails removed. Avg	2	......	55.7	35	9.3	8.3	.5	......	.5	175
Hake, entrails removed	1	......	52.5	39.5	8	7.2	.3	......	.5	145
Cusk, entrails removed	1	......	40.3	49	10.7	10.1	.1	......	.5	190
Haddock, entrails removed. Min	4	......	48	38.5	8.6	7.8	.1	......	.5	150
Haddock, entrails removed. Max	4	......	52.9	42.9	9.6	8.9	.2	......	.8	170
Haddock, entrails removed. Avg	4	......	51	40	9	8.2	.2	......	.6	160
Cod, whole. Min	2	......	48.5	35.1	8.4	7.7	.1	......	.6	145
Cod, whole. Max	2	......	56.5	42.3	9.2	8.3	.3	......	.6	165
Cod, whole. Avg	2	......	52.1	38.7	8.8	8	.2	......	.6	155
Cod, dressed. Min	3	......	25.5	55.3	11	9.9	.2	......	.8	190
Cod, dressed. Max	3	......	33.7	62.1	12.4	11.4	.3	......	.9	220
Cod, dressed. Avg	3	......	29.9	58.5	11.6	10.6	.2	......	.8	205
Tomcod, whole	1	......	59.9	32.7	7.4	6.8	.2	......	.4	135
Pollock, dressed	1	......	28.5	54.3	17.2	15.5	.6	......	1.1	315
Halibut, sections. Min	3	......	11.2	60.9	16	13.4	1.7	......	.7	320
Halibut, sections. Max	3	......	23.1	62.6	26.5	16.1	9.4	......	1	695
Halibut, sections. Avg	3	......	17.7	61.9	20.4	15.1	4.4	......	.9	465
Turbot, whole	1	......	47.7	37.3	15	6.8	7.5	......	.7	440
Flounder, whole. Min	2	......	56.2	35.8	6.8	6.1	.2	......	.5	120
Flounder, whole. Max	2	......	57	37	7.2	6.3	.3	......	.6	130
Flounder, whole. Avg	2	......	56.6	36.4	7	6.2	.3	......	.5	125
Flounder, entrails removed	1	......	57	35.8	7.2	6.3	.3	......	.6	130
Lamprey eel, whole	1	......	45.8	38.5	15.7	8.1	7.2	......	.4	455
Skate, lobe of body	1	......	51	40.2	8.8	7.5	.7	......	.6	170
Preserved fish:										
Salted mackerel	1	7.1	33.3	28.1	31.5	14.7	15.1	......	1.7	910
Salted cod. Min	2	17.2	24.3	40	17.2	15.7	.3	......	1.2	305
Salted cod. Max	2	17.3	25.5	40.5	18	16.4	.4	......	1.2	320
Salted cod. Avg	2	17.2	24.9	40.3	17.6	16	.4	......	1.2	315
Smoked herring	1	6.5	44.4	19.2	29.9	20.2	8.8	......	.9	745
Smoked haddock	1	1.4	32.2	49.2	17.2	16.1	.1	......	1	305
Smoked halibut. Min	2	12	5.9	44.9	33	16.7	13.6	......	1.9	920
Smoked halibut. Max	2	12.1	8	47	37.1	21.6	14.4	......	1.9	975
Smoked halibut. Avg	2	12.1	6.9	46	35	19.1	14	......	1.9	945

TABLE 1.—*Composition of food materials as found in the markets, etc.*—Continued.

Food materials.	Number of analyses.	Salt.	Refuse (bones, skin, shell, etc.).	Edible portion.						
				Water.	Nutrients.					Fuel value of 1 pound.
					Total.	Protein.	Fat.	Carbohydrates.	Mineral matters.	
FISH, SHELLFISH, ETC.—cont'd.										
Shellfish, etc.:		*P. ct.*	*P. ct.*	*P. ct.*	*P. ct.*	*P. ct.*	*P. ct.*	*P. ct.*	*P. ct.*	*Calories.*
Oysters in shell... Min..	34	......	76.1	9.2	1	.5	.1	.2	.1	......
Oysters in shell... Max..	34	......	88.8	22.8	4.6	2.1	.4	1.3	.7	......
Oysters in shell... Avg..	34	......	82.4	15.3	2.3	1.1	.2	.6	.4	40
Long clams in shell Min..	4	......	42.1	46.4	7.5	4.4	.5	.9	1.2	120
Long clams in shell Max..	4	......	46.1	49.9	8.5	5.1	.7	1.4	1.6	145
Long clams in shell Avg..	4	......	43.6	48.4	8	4.8	.6	1.1	1.5	135
Round clams in shell	1	......	68.3	27.3	4.4	2.1	.1	1.3	.9	65
Mussels in shell	1	......	49.3	42.7	8	4.4	.6	2.1	.9	145
Lobster in shell... Min..	4	......	47.5	25.1	5.5	4.1	.5	......	.6	125
Lobster in shell... Max..	4	......	69.1	44.3	8.2	6.5	.9	......	.8	160
Lobster in shell... Avg..	4	......	62.1	31.1	6.8	5.4	.7	......	.7	130
Crab in shell	1	......	55.8	34.1	10.1	7.9	.9	......	1.3	185
Terrapin in shell	1	......	70	15.7	5.3	4.4	.7	......	.2	110
Green turtle in shell.......	1	......	76	19.2	4.8	4.4	.1	......	.3	85

TABLE 2.—*Composition of food materials, edible portion.*

Food materials.	Number of analyses.	Salt.	Water.	Nutrients.					Fuel value of 1 pound.
				Total.	Protein.	Fat.	Carbohydrates.	Mineral matters.	
MEATS, ETC.									
Beef:			*P. ct.*	*P. ct.*	*P. ct.*	*P. ct.*	*P. ct.*	*P. ct.*	*Calories*
Neck............... Min..	3	....	60.6	35.5	18.3	14.5	....	1	985
Neck............... Max..	3	....	64.5	39.4	20.2	20.1	....	1.1	1,195
Neck............... Avg..	3	....	62	38	19.5	17.5	....	1	1,100
Chuck ribs .... Min..	8	....	51.3	33.8	16	14.7	....	.7	960
Chuck ribs .... Max..	8	....	66.2	48.7	20.4	32	....	1.2	1,650
Chuck ribs .... Avg..	8	....	58	42	17.6	23.5	....	.9	1,320
Ribs............... Min..	5	....	46.2	50.1	14.6	32.9	....	.8	1,690
Ribs............... Max..	5	....	49.9	53.8	16.1	37.1	....	1.1	1,860
Ribs............... Avg..	5	....	48.1	51.9	15.4	35.6	....	.9	1,790
Brisket........................	1	....	47.4	52.6	14.6	37.2	....	.8	1,840
Cross ribs.....................	1	....	43.9	56.1	13.7	41.6	....	.8	2,010
Shoulder........... Min..	6	....	56.2	30.8	17.1	10	....	.9	790
Shoulder........... Max..	6	....	69.2	43.8	21	21.6	....	1.2	1,300
Shoulder........... Avg..	6	....	63.9	36.1	19.5	15.6	....	1	1,020
Shin...........................	1	....	73.8	26.2	22.7	2.3	....	1.2	520
Plate..........................	1	....	44.4	55.6	15.4	39.4	....	.8	1,950
Navel..........................	1	....	47.6	52.4	15.1	36.5	....	.8	1,820
Sirloin............ Min..	4	....	58.9	39.3	16.9	16.1	....	.9	1,090
Sirloin............ Max..	4	....	60.7	41.1	22	22.9	....	1.4	1,290
Sirloin............ Avg..	4	....	60	40	18.5	20.5	....	1	1,210
Socket	1	....	57.1	42.9	16.7	25.2	....	1	1,375
Rump .... Min..	2	....	40.2	43.7	14.7	26.8	....	.8	1,430
Rump .... Max..	2	....	56.3	59.8	16.1	41.3	....	.8	2,145
Rump .... Avg..	2	....	48.2	51.8	15.4	35.6	....	.8	1,790
Round .... Min..	3	....	66	30.5	19.5	8.3	....	1.1	745
Round .... Max..	3	....	69.5	34	21.5	13.4	....	1.3	935
Round .... Avg..	3	....	68.2	31.8	20.5	10.1	....	1.2	805
Leg	1	....	72.1	27.9	21	5.7	....	1.2	630
Top of sirloin	1	....	42.2	57.8	13.3	43.7	....	.8	2,090
Flank	1	....	27.4	72.6	12	59.9	....	.7	2,750
Fore quarter	....	....	54.1	45.9	17.3	27.7	....	.9	1,495
Hind quarter...................	....	....	55.7	44.3	17.1	26.3	....	.9	1,430
Side without kidney fat........	....	....	54.8	45.2	17.2	27.1	....	.9	1,465
Liver..........................	1	....	69.5	30.5	20.1	5.4	3.5	1.5	665
Kidney.........................	1	....	75.7	24.3	17	4.8	1.3	1.2	545
Heart	1	....	56.5	43.5	16.3	26.2	....	1	1,410
Tongue	1	....	63.5	36.5	17.4	18	....	1.1	1,085
Kidney fat	1	....	4.3	95.7	.9	94.6	....	.2	4,010
Marrow (leg bone)	1	....	3.3	96.7	2.6	92.8	....	1.3	3,965
Veal:									
Shoulder .... Min..	2	....	65.6	28	19.7	6.1	....	1.2	640
Shoulder .... Max..	2	....	72	34.4	20.7	13.5	....	1.2	935
Shoulder .... Avg..	2	....	68.8	31.2	20.2	9.8	....	1.2	790

TABLE 2.—*Composition of food materials, edible portion*—Continued.

Food materials.	Number of analyses.	Salt.	Water.	Nutrients. Total.	Protein.	Fat.	Carbohydrates	Mineral matters.	Fuel value of 1 pound.
MEATS, ETC.—continued.									
Mutton:			*P. ct.*	*P. ct.*	*P. ct.*	*P. ct.*	*P. ct.*	*P. ct.*	*Calories*
Shoulder	1	......	58.6	41.4	18.1	22.4	......	.9	1.280
Breast	1	......	37.6	62.4	14.2	47.2	......	1	2.255
Rack	1	......	54.9	45.1	18.4	25.9	......	.8	14,85
Neck	1	......	55.7	44.3	16.2	27.3	......	.8	1,455
Leg	1	......	61.8	38.2	18.3	19	......	.9	1,140
Loin	1	......	49.3	50.7	15	35	......	.7	1,755
Flank	1	......	38.7	61.3	15.8	45	......	.5	2,195
Fore quarter		......	55.2	47.8	17	29.9	......	.9	1.580
Hind quarter		......	54.7	45.3	16.9	27.5	......	.9	1,475
Side without kidney fat		......	53.5	46.5	16.9	28.7	......	.9	1,525
Lamb:									
Shoulder	1	......	51.8	48.2	17.5	29.7	......	1	1,580
Breast	1	......	56.2	43.8	19.2	23.6	......	1	1,355
Neck	1	......	56.7	43.3	17.5	24.8	......	1	1,375
Leg	1	......	64.7	35.3	18.9	15.3	......	1.1	1,000
Loin	1	......	54.8	45.2	19	25.1	......	1.1	1,410
Fore quarter		......	55.1	44.9	18.1	25.8	......	1	1,425
Hind quarter		......	60.9	39.1	18.9	19.1	......	1.1	1,155
Side without kidney fat		......	57.9	42.1	18.6	22.5	......	1	1,295
Liver	1	......	52.7	47.3	24.2	13.2	7.9	2	1,155
Heart	1	......	67.4	32.6	18.3	13.4	......	.9	905
Lungs	1	......	74.6	25.4	21.5	2.6	......	1.5	510
Pork:									
Shoulder roast — Min	4	......	45.8	46.9	14.9	30.4	......	.9	1.590
Shoulder roast — Max	4	......	53.1	54.2	16.9	37.7	......	.9	1,880
Shoulder roast — Avg	4	......	50.3	49.7	16	32.8	......	.9	1,630
Poultry, etc.:									
Chicken	1	......	72.2	27.8	24.4	2	......	1.4	540
Chicken liver	1	......	69.3	30.7	22.3	4.2	2.4	1.8	635
Chicken heart	1	......	72	28	21.2	5.4	......	1.4	620
Chicken gizzard	1	......	72.5	27.5	24.7	1.4	......	1.4	520
Turkey	1	......	66.2	33.8	23.9	8.7	......	1.2	810
Turkey liver	1	......	69.6	30.4	22.9	5.2	.6	1.7	655
Turkey heart	1	......	68.6	31.4	17.2	13.2	......	1	875
Turkey gizzard	1	......	62.7	37.3	21.7	14.5	......	1.1	1,015
Hens' eggs in shell — Min	27	......	72	24.8	13.8	9.1	......	.7	650
Hens' eggs in shell — Max	27	......	75.3	28	16.1	12.5	......	1.6	795
Hens' eggs in shell — Avg	27	......	73.8	26.2	14.9	10.5	......	.8	720
Preserved meats:									
Corned beef, rump	1	......	58.1	41.9	13.3	26.6	......	2	1,370
Corned beef, flank — Min	2	......	43.2	43.5	12.9	24.9	......	2.8	1,340
Corned beef, flank — Max	2	......	56.5	56.8	15.5	41.1	......	3.1	1.975
Corned beef, flank — Avg	2	......	49.8	50.2	14.2	33	......	3	1,655
Corned beef, canned — Min	2	......	51.9	40.4	24.5	14.1	......	3.4	1.130
Corned beef, canned — Max	2	......	53.0	48.1	28.9	20.2	......	3.4	1,310
Corned beef, canned — Avg	2	......	52.8	47.2	26.7	17.1	......	3.4	1,220
Dried beef — Min	2	......	58	40.8	27.6	4.2	.2	6.3	740
Dried beef — Max	2	......	59.2	42	29.9	4.6	2.7	7.3	755
Dried beef — Avg	2	......	58.6	41.4	28.8	4.4	1.4	6.8	745
Tripe, soused	1	......	84	16	13.9	1.8	......	.3	335
Salt pork, fat	1	......	12.1	87.9	.9	82.8	......	4.2	3,510
Smoked ham	1	......	41.5	58.5	16.7	39.1	......	2.7	1,960
Pork sausage — Min	2	......	37.8	55.5	13.5	40.1	......	1.9	1.945
Pork sausage — Max	2	......	44.5	62.2	14.1	45.5	......	2.6	2,180
Pork sausage — Avg	2	......	41.2	58.8	13.8	42.8	......	2.2	2,065
Bologna sausage	1	......	62.4	37.6	18.8	15.8	......	3	1,015
FISH, SHELLFISH, ETC.									
Fresh fish:									
Sturgeon	1	......	78.7	21.3	18	1.9	......	1.4	415
Red horse	1	......	78.6	21.4	17.9	2.3	......	1.2	430
Herring	1	......	69	31	18.5	11	......	1.5	810
Alewife — Min	2	......	72.7	24.1	18.8	3.8	......	1.5	510
Alewife — Max	2	......	15.9	27	19.5	6	......	1.5	615
Alewife — Avg	2	......	74.4	25.6	19.2	4.9	......	1.5	565
Shad — Min	7	......	65.3	26.4	17.8	6.5	......	.9	630
Shad — Max	7	......	73.6	34.8	20	13.6	......	1.5	940
Shad — Avg	7	......	70.6	29.4	18.6	9.5	......	1.3	745
Smelt — Min	2	......	78.2	19.8	15.9	1.7	......	1.4	375
Smelt — Max	2	......	80.2	21.8	18.8	1.9	......	2	420
Smelt — Avg	2	......	79.2	20.8	17.3	1.8	......	1.7	400
Whitefish	1	......	69.8	30.2	22.1	6.5	......	1.6	685
Ciscoe	1	......	76.1	23.9	19.1	3.0	......	1.2	505

TABLE 2.—*Composition of food materials, edible portion*—Continued.

Food materials.		Number of analyses.	Salt.	Water.	Nutrients. Total.	Protein.	Fat.	Carbohydrates.	Mineral matters.	Fuel value of 1 pound.
FISH, SHELLFISH, ETC.—cont'd.										
Fresh fish—Continued.				*P. ct.*	*P. ct.*	*P. ct.*	*P. ct.*	*P. ct.*	*P. ct.*	*Calories*
	Min..	2	....	62.7	35.5	17	16.5	....	1	1,030
California salmon	Max..	2	....	64.5	37.3	18	19.3	....	1.1	1,130
	Avg..	2	....	63.6	36.4	17.4	17.9	....	1.1	1,080
	Min..	5	....	61	32.9	19.2	12.5	....	1.2	885
Salmon	Max..	5	....	67.2	39	24.5	15	....	1.6	1,010
	Avg..	5	....	63.6	36.4	21.6	13.4	....	1.4	965
	Min..	2	....	68.8	30.5	17.3	10.2	....	1.2	785
Lake trout	Max..	2	....	69.5	31.2	19.1	12.6	....	1.4	855
	Avg..	2	....	69.1	30.9	18.2	11.4	....	1.3	820
	Min..	3	....	75.8	20.2	18.5	.8	....	1	380
Brook trout	Max..	3	....	79.8	21.2	20	2.9	....	1.4	495
	Avg..	3	....	77.7	22.3	19	2.1	....	1.2	440
	Min..	2	....	79.5	20.2	18.4	.5	....	1.1	365
Pickerel	Max..	2	....	79.8	20.5	18.9	.5	....	1.2	375
	Avg..	2	....	79.7	20.3	18.6	.5	....	1.2	365
Pickerel, pike		1	....	79.8	20.2	18.6	.6	....	1	370
Muscalonge		1	....	76.3	23.7	19.6	2.5	....	1.6	470
	Min..	2	....	69.8	26.6	17.6	7.9	....	.9	960
Eel, salt water	Max..	2	....	73.4	30.2	19	10.3	....	1.1	790
	Avg..	2	....	71.6	28.4	19.3	9.1	....	1	725
Mullet		1	....	74.9	25.1	19.3	4.6	....	1.2	555
	Min..	6	....	64	21.3	17.5	2.2	....	1	430
Mackerel	Max..	6	....	78.7	36	19.3	16.3	....	1.5	1,025
	Avg..	6	....	73.4	26.6	18.2	7.1	....	1.3	640
Spanish mackerel		1	....	68.1	31.9	20.6	9.8	....	1.5	790
	Min..	2	....	67.4	21.8	18.2	1.6	....	1	425
Pompano	Max..	2	....	78.2	32.6	19.2	13.5	....	1	910
	Avg..	2	....	72.8	27.2	18.6	7.6	....	1	665
Bluefish		1	....	78.5	21.5	19	1.2	....	1.3	405
Butter-fish		1	....	70	30	17.8	11	....	1.2	795
	Min..	2	....	74.8	21.4	19.2	1	....	1.2	400
Black bass	Max..	2	....	78.6	25.2	21.5	2.5	....	1.2	505
	Avg..	2	....	76.7	23.3	20.4	1.7	....	1.2	450
	Min..	2	....	78.1	19.6	17.9	.6	....	1.1	360
Yellow perch	Max..	2	....	80.4	21.9	19.5	1.1	....	1.3	410
	Avg..	2	....	79.3	20.7	18.7	.8	....	1.2	380
Wall eyed pike		1	....	79.7	20.3	18.4	.5	....	1.4	365
Gray pike		1	....	80.8	19.2	17.3	.8	....	1.1	355
	Min..	6	....	75.8	20.3	16.7	1.6	....	.9	405
Striped bass	Max..	6	....	79.7	24.2	18.8	4.6	....	1.4	525
	Avg..	6	....	77.7	22.3	18.3	2.8	....	1.2	460
	Min..	2	....	75.6	24.2	17.6	2.5	....	1.1	485
White perch	Max..	2	....	75.8	24.4	20.4	5.6	....	1.3	565
	Avg..	2	....	75.7	24.3	19	4.1	....	1.2	525
Sea bass		1	....	79.3	20.7	18.8	.5	....	1.4	370
	Min..	2	....	79	20	18.5	.5	....	1.1	365
Grouper	Max..	2	....	80	21	19.2	.7	....	1.1	385
	Avg..	2	....	79.4	20.6	18.9	.6	....	1.1	375
	Min..	3	....	77.3	20.2	18.3	.5	....	1.3	360
Red snapper	Max..	3	....	79.8	22.7	19.9	1.9	....	1.3	440
	Avg..	3	....	78.5	21.5	19.2	1	....	1.3	400
	Min..	3	....	72	20.3	17.5	1.5	....	1.4	390
Porgy	Max..	3	....	79.7	28	19.3	7.9	....	1.4	685
	Avg..	3	....	75	25	18.5	5.1	....	1.4	560
	Min..	2	....	72	20.9	18.9	.7	....	1.1	380
Sheepshead	Max..	2	....	79.1	28	20.2	6.7	....	1.3	600
	Avg..	2	....	75.6	24.4	19.5	3.7	....	1.2	520
Red bass		1	....	81.6	18.4	16.7	.5	....	1.2	330
Kingfish		1	....	79.2	20.8	18.7	.9	....	1.2	385
Weakfish		1	....	79	21	17.4	2.4	....	1.2	425
	Min..	4	....	77	18.6	17.4	.6	....	.7	350
Blackfish	Max..	4	....	81	23	19	2.8	....	1.4	470
	Avg..	4	....	79.1	20.9	18.5	1.3	....	1.1	400
Hake		1	....	83.1	16.9	15.2	.7	....	1	310
Cusk		1	....	82	18	16.9	.2	....	.9	325
	Min..	4	....	80.3	17.4	15.9	.1	....	1	305
Haddock	Max..	4	....	82.6	19.7	18.4	.4	....	1.6	350
	Avg..	4	....	81.7	18.3	16.8	.3	....	1.2	325
	Min..	5	....	80.7	16.5	15	.3	....	1	285
Cod	Max	5	....	83.5	19.3	17.6	.5	....	1.3	340
	Avg..	5	....	82.6	17.4	15.8	.4	....	1.2	310
Tomcod		1	....	81.5	18.5	17.1	.4	....	1	335
Pollock		1	....	76	24	21.7	.8	....	1.5	410
	Min..	3	....	79.1	20.8	17.5	2.2	....	.9	420
Halibut	Max..	3	....	79.2	29.9	19.4	10.6	....	1.2	785
	Avg..	3	....	75.4	24.6	18.3	5.2	....	1.1	560

TABLE 2.—*Composition of food materials, edible portion*—Continued.

Food materials.		Number of analyses.	Salt.	Water.	Nutrients. Total.	Protein.	Fat.	Carbohydrates.	Mineral matters.	Fuel value of 1 pound.
FISH, SHELLFISH, ETC.—cont'd.										
Fresh fish—Continued.				*P. ct.*	*P. ct.*	*P. ct.*	*P. ct.*	*P. ct.*	*P. ct.*	*Calories*
Turbot		1	....	71.4	28.6	12.9	14.4	....	1.3	850
Flounder	Min	3	....	83.4	15	12.9	.4	....	1.2	275
	Max	3	....	85	16.6	14.7	.7	....	1.3	300
	Avg	3	....	84.2	15.8	13.9	.6	....	1.3	285
Lamprey eel		1	....	71.1	28.9	14.9	13.3	....	.7	840
Skate		1	....	82.2	17.8	15.3	1.4	....	1.1	345
Preserved fish:										
Desiccated cod		1	2.9	15.2	81.9	74.6	1.9	....	5.4	1,470
Salt cod	Min	2	22.7	53.5	23	21.2	.3	....	1.6	405
	Max	2	23.4	53.6	23.8	21.7	.4	....	1.6	420
	Avg	2	23	53.6	23.4	21.4	.4	....	1.6	410
Boned cod		1	21.5	54.3	24.2	22.2	.3	....	1.7	425
Salt mackerel		1	10.6	42.2	47.2	22.1	22.6	....	2.5	1,365
Smoked haddock		11	.2	72.5	25.4	23.7	.2	....	1.5	450
Smoked halibut	Min	2	12.9	47.7	35.9	18.2	14.4	....	2	905
	Max	2	13.1	51.1	39.4	23	15.6	....	2.1	1,035
	Avg	2	12.9	49.4	37.7	20.6	15.1	....	2	1,020
Canned mackerel		1	1.9	68.2	29.9	19.9	8.7	....	1.3	735
Canned salmon	Min	3	.4	57.6	33.6	19.2	11.1	....	1.3	865
	Max	3	2.2	65.9	42	21.3	21.5	....	1.4	1,265
	Avg	3	1	61.9	37.1	20.1	15.7	....	1.3	1,035
Canned sardines		1	....	56.4	43.6	25.3	12.7	....	5.6	1,005
Canned tunny		1	....	72.7	27.3	21.5	4.1	....	1.7	575
Canned salt mackerel.	Min	2	9.4	43.2	45.2	16.9	24.8	....	2.5	1,375
	Max	2	11.2	43.6	47.3	17.7	27.9	....	2.7	1,490
	Avg	2	10.3	43.4	46.3	17.3	26.3	....	2.6	1,430
Canned smoked haddock		1	5.6	68.7	25.7	21.8	3.3	....	1.6	505
Shellfish, etc.:										
Oysters in shell	Min	34	....	81.7	8.6	4.2	.6	1.8	1.2	135
	Max	34	....	91.4	18.3	8.5	1.7	6.7	2.8	345
	Avg	34	....	87.1	12.9	6.1	1.2	3.6	2	230
Oysters, "solids"	Min	4	....	85.2	11.6	5.9	1.5	3.2	.8	230
	Max	4	....	88.4	14.8	6.6	1.8	5.6	1.1	300
	Avg	4	....	87.2	12.8	6.3	1.6	4	.9	260
Canned oysters	Min	3	....	84.6	14	7	2	4.1	1.3	285
	Max	3	....	86	15.4	8	2.2	5.2	1.4	310
	Avg	3	....	85.3	14.7	7.4	2.1	3.9	1.3	300
Long clams from shell.	Min	4	....	85	13.9	8.1	1	1.5	2.1	225
	Max	4	....	86.1	15	9	1.2	2.5	3	255
	Avg	4	....	85.8	14.2	8.6	1	2	2.6	240
Long clams, canned		1	....	84.5	15.5	9	1.3	2.9	2.3	275
Round clams from shell		1	....	86.2	13.8	6.5	.4	4.2	2.7	215
Round clams, canned		1	....	83	17	10.4	.8	3	2.8	285
Scallops	Min	2	....	77.8	17.2	14.4	....	1.1	1.3	310
	Max	2	....	82.8	22.2	15.1	.3	5.7	1.5	385
	Avg	2	....	80.3	19.7	14.7	.2	3.4	1.4	345
Mussels from shell		1	....	84.2	15.8	8.7	1.1	4.1	1.9	285
Lobster from shell	Min	4	....	79.2	15.7	12.3	1.5	....	1.6	310
	Max	4	....	84.3	20.8	17.8	2.5	....	1.9	395
	Avg	4	....	81.8	18.2	14.6	1.9	....	1.7	350
Lobster, canned	Min	2	....	76.2	20.6	17.4	.5	....	2.2	340
	Max	2	....	79.4	23.8	20	1.7	....	2.8	445
	Avg	2	....	77.7	22.3	18.7	1.1	....	2.5	395
Crayfish		1	....	81.2	18.8	17	.5	....	1.3	335
Crab		1	....	77.1	22.9	17.8	2	....	3.1	415
Crabs, canned	Min	2	....	79	19	16.5	.8	....	1.8	340
	Max	2	....	81	21	16.7	2.3	....	2.1	410
	Avg	2	....	80	20	16.5	1.5	....	2	370
Shrimp		1	....	70.8	29.2	25.6	1	....	2.6	520
Terrapin		1	....	74.5	25.5	21	3.5	....	1	540
Green turtle		1	....	79.8	20.2	18.5	.5	....	1.2	365
DAIRY PRODUCTS, ETC.										
Milk		....	....	87	13	3.6	4	4.7	.7	325
Butter		....	....	10.5	89.5	1	85	.5	3	3,615
Cheese, full cream	Min	5	....	27	63.2	26	30	.8	3.7	1,815
	Max	5	....	36.8	73	30.6	38.3	3.5	4.8	2,185
	Avg	5	....	30.2	69.8	28.3	35.5	1.8	4.2	2,070
Cheese, skim-milk		1	....	41.3	58.7	38.4	6.8	8.9	4.6	1,165
Oleomargarine		....	....	11	89	.6	85	.4	3	3,605

TABLE 2.—*Composition of food materials, edible portion*—Continued.

Food materials.		Number of analyses.	Salt.	Water.	Nutrients: Total.	Protein.	Fat.	Carbohydrates.	Mineral matters.	Fuel value of 1 pound.
VEGETABLE FOODS.				*P. ct.*	*P. ct.*	*P. ct.*	*P. ct.*	*P. ct.*	*P. ct.*	*Calories*
Potatoes[1]	Min	12	....	75.4	17.8	1.1	....	14.3	.8	315
	Max	12	....	82.2	24.6	3	.1	21.2	1.2	445
	Avg	12	....	78.9	21.1	2.1	.1	17.9	1	375
Sweet potatoes[1]	Min	6	....	66	25.6	.5	.3	18.6	.7	465
	Max	6	....	74.4	34	3.6	.6	32.2	1.3	620
	Avg	6	....	71.1	28.9	1.5	.4	26	1	530
Red beets[1]	Min	9	....	85.5	7.8	1.1	.1	4.5	.7	130
	Max	9	....	92.2	14.5	1.8	.3	13	1.6	250
	Avg	9	....	88.5	11.5	1.5	.1	8.8	1.1	195
Turnips[1]	Min	7	....	87.1	7.6	.8	.1	5	.7	135
	Max	7	....	92.4	12.9	1.4	.3	10.5	1.4	225
	Avg	7	....	89.4	10.6	1.2	.2	8.2	1	185
Carrots[1]	Min	8	....	86.5	8.9	.8	.2	6	.6	155
	Max	8	....	91.1	13.5	2	.7	12.7	1.3	235
	Avg	8	....	88.6	11.4	1.1	.4	8.9	1	205
Onions[1]	Min	6	....	81.5	6.5	.8	.2	4.2	.4	120
	Max	6	....	93.5	18.5	2.3	.4	15.5	.7	335
	Avg	6	....	87.6	12.4	1.4	.3	10.1	.6	225
Squash, flesh[1]	Min	3	....	85.3	10.3	.7	....	8.1	.5	190
	Max	3	....	89.7	14.7	1.1	.3	13.2	.9	265
	Avg	3	....	88.1	11.9	.9	.2	10.1	.7	215
Pumpkin, flesh[1]	Min	2	....	92.4	5.6	.9	.1	3.9	.6	95
	Max	2	....	94.4	7.6	1	.1	5.9	.7	130
	Avg	2	....	93.4	6.6	.9	.1	4.9	.7	110
Cucumber[1]	Min	2	....	95.7	3.7	.8	.2	2.3	.5	65
	Max	2	....	96.3	4.3	.8	.2	2.8	.5	75
	Avg	2	....	96	4	.8	.2	2.5	.5	70
Cabbage, entire	Min	2	....	87.5	6.4	2.1	.2	3.4	.7	110
	Max	2	....	93.6	12.5	2.7	.5	7.2	2.1	205
	Avg	2	....	90.5	9.5	2.4	.4	5.3	1.4	155
Cabbage, inner leaves[1]	Min	2	....	91.9	5.7	1.5	.2	3.4	.6	100
	Max	2	....	94.3	8.1	1.5	.2	5.7	.6	140
	Avg	2	....	93.1	6.9	1.5	.2	4.6	.6	120
Cauliflower		1	....	90.8	9.2	1.6	.8	5	.8	155
Lettuce	Min	3	....	91.5	5.4	1.4	.4	2.4	.9	90
	Max	3	....	94.6	8.5	1.8	.6	4.9	1.2	150
	Avg	3	....	93.1	6.9	1.6	.5	3.7	1.1	120
Spinach		1	....	92.4	7.6	2.1	.5	3.1	1.9	120
Rhubarb, stems		1	....	92.7	7.3	.8	1.2	4.4	.9	145
Asparagus	Min	3	....	93.6	5.8	1.6	.2	3.1	.5	105
	Max	3	....	94.2	6.4	2.1	.3	3.5	1	110
	Avg	3	....	94	6	1.8	.2	3.3	.7	105
Tomatoes	Min	6	....	95.5	3.7	.7	.3	2.2	.3	75
	Max	6	....	96.3	4.5	.9	.5	2.8	.4	80
	Avg	6	....	96	4	.8	.4	2.5	.3	80
Green peas		1	....	78.1	21.9	4.4	.5	16.1	.9	400
String beans	Min	2	....	83.5	9	1.7	.3	6.3	.7	160
	Max	2	....	91	16.5	2.8	.4	12.6	.8	305
	Avg	2	....	87.2	12.8	2.2	.4	9.5	.7	235
Lima beans, green		1	....	68.5	31.5	7.1	.7	22	1.7	370
Okra		1	....	87.4	12.6	2	.4	9.5	.7	230
Green sweet c(rn		1	....	81.2	18.8	2.8	1.1	14.2	.7	360
E: gplant		1	....	92.9	7.1	1.2	.3	5.1	.5	130
Peas		1	....	83.9	16.1	.6	.8	14.2	.5	290
Peas, canned	Min	82	....	77.5	7.3	1.6	.1	6.1	.3	130
	Max	82	....	92.7	22.5	6.1	.4	15.1	2	405
	Avg	82	....	85.4	14.6	3.6	.2	9.7	1.1	255
Haricots verts, canned	Min	6	....	94.3	3.9	.9	....	2	.9	55
	Max	6	....	96.1	5.7	1.4	.3	3.1	1.3	95
	Avg	6	....	95.1	4.9	1.1	.1	2.6	1.1	70
String beans, canned	Min	18	....	94.7	3.7	.6	....	2	.5	60
	Max	18	....	96.3	9.3	1.5	.1	5.9	2.3	140
	Avg	18	....	94.3	5.7	.9	.1	3.5	1.2	85
Stringless beans, canned	Min	7	....	92.4	5.6	.1	.1	2.8	.9	65
	Max	7	....	94.4	7.6	1.3	.1	4.8	1.8	115
	Avg	7	....	93.9	6.1	1.1	.1	3.5	1.4	90
Haricots flageolets, canned.	Min	3	....	80.4	16.1	4	....	10.8	.9	275
	Max	3	....	83.9	19.6	5.2	.1	13.5	1.7	335
	Avg	3	....	81.6	18.4	4.6	.1	12.5	1.2	320

[1] These, as ordinarily found in the market, have more or less refuse. The figures for composition for all of them except "cabbage, entire," as for the other vegetable food materials apply to the edible portion. Observations made in connection with studies of dietaries here have led to the use of the following figures as representing the percentages of refuse: Onions, 10 per cent; sweet potatoes, 12.5 per cent; potatoes, 15 per cent; cabbage, 15.5 per cent; turnips, 30 per cent; squash, 50 per cent; apples and grapes, each 25 per cent.

TABLE 2.—*Composition of food materials, edible portion*—Continued.

Food materials.		Number of analyses.	Salt.	Water.	Nutrients. Total.	Protein.	Fat.	Carbohydrates.	Mineral matters.	Fuel value of 1 pound.
VEGETABLE FOODS—continued.										
				P. ct.	*P. ct.*	*P. ct.*	*P. ct.*	*P. ct.*	*P. ct.*	*Calories*
Haricots panaches, canned		1	....	86.1	13.9	3.7	....	9.2	1	210
Little green beans, canned		1	....	93.8	6.2	1.2	.1	3.4	1.5	90
Wax beans, canned		1	....	94.7	5.3	1	.1	3	1.2	80
Lima beans, canned	Min	15	....	75.7	16.1	3.2	.2	10.5	1	275
	Max	15	....	83.9	24.3	5.7	.6	17.0	2.6	445
	Avg	15	....	79.7	20.3	4	.3	14.4	1.0	355
Baked beans, canned	Min	12	....	64.3	30.6	6.4	1.3	19.1	1.7	595
	Max	12	....	69.4	35.7	7.7	5.5	21.2	2.6	745
	Avg	12	....	67.2	32.8	7.1	3.2	20.3	2.2	645
Red kidney beans, canned		1	....	72.7	27.3	7	.2	18.5	1.6	485
Corn, canned	Min	43	....	68.3	16.3	2	.7	13	.5	310
	Max	43	....	83.7	31.7	3.4	1.9	25.8	1.5	615
	Avg	43	....	75.4	24.6	2.8	1.3	19.6	9	470
Artichokes, canned	Min	3	....	90.2	6.1	.5	....	3.7	1.4	90
	Max	3	....	93.9	9.8	1	....	6.8	2.2	140
	Avg	3	....	92.5	7.5	.8	....	5	1.7	110
Sweet potato		1	....	68.4	31.6	1.3	.3	29.2	.8	580
Okra, canned	Min	4	....	94	5.1	.5	....	3.3	.3	75
	Max	4	....	94.9	6	.9	.2	3.7	1.7	05
	Avg	4	....	94.4	5.6	.7	.1	3.6	1.2	85
Brussels sprouts, canned		1	....	93.8	6.2	1.5	1	3.4	1.3	95
Tomatoes, canned	Min	11	....	92.5	5.1	1.1	.1	3.5	.5	90
	Max	11	....	94.9	7.5	1.6	.3	5.1	1.2	135
	Avg	11	....	93.7	6.3	1.3	.2	4.2	.6	110
Asparagus, canned	Min	14	....	92.9	4.6	.9	....	2.1	.8	70
	Max	14	....	95.4	7.1	2.4	.2	4.1	1.8	115
	Avg	14	....	94.4	5.6	1.5	.1	2.8	1.2	85
Pumpkin, canned	Min	5	....	89.4	5.7	.5	.1	4.7	.4	100
	Max	5	....	94.3	10.6	.9	.4	8.8	.6	195
	Avg	5	....	92.7	7.3	.7	.1	6	.5	130
Squash, canned	Min	2	....	85.6	12.5	.2	.1	10.6	.2	230
	Max	2	....	87.5	14.4	.9	.5	13.9	.6	265
	Avg	2	....	86.6	13.4	.5	.3	12.2	.4	250
Macedoine, canned	Min	5	....	91.5	4.1	.6	....	2.3	.8	56
	Max	5	....	95.9	8.5	1.7	....	5.7	1.2	140
	Avg	5	....	93.1	6.9	1.3	....	4.6	1	110
Succotash, canned	Min	10	....	71.4	20.1	2.9	.7	15	.7	375
	Max	10	....	79.9	28.6	4.1	1.1	22.4	1.5	530
	Avg	10	....	76.2	23.8	3.5	.9	18.5	.9	445
Mixed corn and tomatoes, canned	Min	2	....	83.6	8.5	1.2	.3	6.4	.5	155
	Max	2	....	91.5	16.4	2.1	.4	12.7	1.2	285
	Avg	2	....	87.5	12.5	1.7	.4	9.5	.9	225
Mixed okra and tomatoes, canned	Min	3	....	91.4	7.7	1.1	.2	5.1	1.4	125
	Max	3	....	92.3	8.3	1.2	.3	5.7	1.8	140
	Avg	3	....	91.8	8.2	1.2	.2	5.2	1.6	130
Apples flesh	Min	2	....	82.2	15.9	.2	.3	15.1	.2	300
	Max	2	....	84.1	17.8	.3	.5	16.7	.3	335
	Avg	2	....	83.2	16.8	.3	.4	15.9	.2	320
Cherries, flesh		1	....	86.1	13.9	1.1	.8	11.4	.6	265
Strawberries	Min	19	....	87.7	6	.6	.4	4.4	.4	120
	Max	19	....	94	12.3	1.2	1.1	8.3	.8	235
	Avg	19	....	90.8	9.2	1	.7	6.9	.6	175
Blackberries		1	....	88.9	11.1	.9	2.1	7.5	.6	245
Whortleberries		1	....	82.4	17.6	.7	3	13.5	.4	390
Cranberries		1	....	87.6	12.4	.4	.9	10.9	.2	350
Grapes, Catawba		1	....	74.8	25.2	1.6	1.7	21.3	.6	500
Lemons	Min	2	....	88.4	9.8	.8	.2	8.2	.5	175
	Max	2	....	90.2	11.6	1.1	1.6	8.5	.5	245
	Avg	2	....	89.3	10.7	1	.9	8.3	.5	210
Banana, pulp		1	....	66.3	33.7	1.4	1.4	29.8	1.1	640
Pineapple		1	....	89.3	10.7	.4	.3	9.7	.3	200
Watermelon, flesh or pulp		1	....	91.9	8.1	.9	.7	6.2	.3	160
Nutmeg melon, flesh or pulp		1	....	76.4	23.6	1.4	.2	20.5	1.5	415
Rice	Min	10	....	11.4	86	.9	.3	77.6	.3	1,600
	Max	10	....	14	88.6	8.6	.6	81	.5	1,655
	Avg	10	....	12.4	87.6	7.4	.4	79.4	.4	1,630
Beans, dried	Min	3	....	10.4	85	20.4	1.4	57.2	2.7	1,560
	Max	3	....	15	89.6	26.6	3.1	60.4	3.7	1,685
	Avg	3	....	12.6	87.4	23.1	2	59.2	3.1	1,615
Maize meal	Min	77	....	8	72.6	7.1	2	60.9	.9	1,405
	Max	77	....	27.4	92	13.9	5.1	77.1	4.1	1,805
	Avg	77	....	15	85	9.2	3.8	70.6	1.4	1,645
White hominy	Min	2	....	13.4	86.4	8.1	.4	77.4	.4	1,610
	Max	2	....	13.6	86.6	8.4	.5	77.5	.5	1,620
	Avg	2	....	13.5	86.5	8.3	.4	77.4	.4	1,620

TABLE 2.—*Composition of food materials, edible portion*—Continued.

Food materials.		Number of analyses.	Salt.	Water.	Nutrients.					Fuel value of 1 pound.
					Total.	Protein.	Fat.	Carbohydrates.	Mineral matters.	
VEGETABLE FOODS—continued.				*P. ct.*	*P. ct.*	*P. ct.*	*P. ct.*	*P. ct.*	*P. ct.*	*Calories*
Oatmeal	Min..	6	....	6.2	91.2	12.9	6.1	67.3	1.8	1,820
	Max..	6	....	8.8	93.8	16.3	8.8	70.1	2.2	1,875
	Avg..	6	....	7.8	92.2	14.7	7.1	68.4	2	1,845
Pearl barley		1	....	11.8	88.2	8.4	.7	78.1	1	1,635
Rye flour	Min..	4	....	12.4	86.4	.6	.8	77.9	.6	1,615
	Max..	4	....	13.6	87.6	7.1	.9	79.5	.8	1,625
	Avg..	4	....	13.1	86.9	6.7	.8	78.7	.7	1,625
Wheat flour	Min..	22	....	8.2	85.7	8.6	.6	71.6	.3	1,625
	Max..	22	....	14.3	91.8	13.6	1.8	79.5	.7	1,680
	Avg..	22	....	12.5	87.5	11	1.1	74.9	.5	1,645
Graham flour	Min..	3	....	12.1	86.3	11.3	1.5	71.6	1.7	1,610
	Max..	3	....	13.7	87.9	12.4	1.9	72	2	1,645
	Avg..	3	....	13.1	86.9	11.7	1.7	71.7	1.8	1,625
Entire wheat flour	Min..	2	....	12.9	86.9	13.1	1.9	69.5	1.4	1,615
	Max..	2	....	13.1	87.1	14.1	2	70.5	1.4	1,660
	Avg..	2	....	13	87	13.6	2	70	1.4	1,640
Cracked wheat	Min..	2	....	9.8	88.9	11.9	1.5	73.9	1.4	1,660
	Max..	2	....	11.1	90.2	12	1.8	75.2	1.4	1,700
	Avg..	2	....	10.4	89.6	11.9	1.7	74.6	1.4	1,680
Buckwheat flour	Min..	4	....	12.8	82.4	4.2	.7	71.6	.7	1,560
	Max..	4	....	17.6	87.2	8.1	1.8	79.6	1.3	1,640
	Avg..	4	....	14.6	85.4	6.9	1.4	76.1	1	1,605
Buckwheat farina		1	....	11.2	88.8	3.3	.3	84.8	.4	1,650
Buckwheat groats		1	....	10.6	89.4	4.8	.6	83.4	.6	1,665
Wheat bread	Min..	5	....	31.2	66.5	8.6	.6	55.2	.6	1,245
	Max..	5	....	33.5	68.8	9.2	2.5	58.5	1.2	1,300
	Avg..	5	....	32.3	67.7	8.8	1.7	56.3	.9	1,280
Graham bread		1	....	34.2	65.8	9.5	1.4	53.3	1.6	1,225
Rye bread		1	....	30	70	8.4	.5	59.7	1.4	1,285
Boston crackers		1	....	8.3	91.7	10.7	9.9	68.7	2.4	1,895
Soda crackers		1	....	8	92	10.3	9.4	70.5	1.8	1,900
Pilot (bread) crackers		1	....	7.9	92.1	12.4	4.4	74.2	1.1	1,795
Oyster crackers		1	....	3.9	96.1	11.3	4.8	77.5	2.5	1,855
Oatmeal crackers		1	....	4.9	95.1	10.4	13.7	69.6	1.4	2,065
Graham crackers		1	....	5	95	9.8	13.6	69.7	1.9	2,050
Starch		2	....	2	98	....	....	97.8	.2	1,820
Sugar, granulated		....	....	2	98	....	....	97.8	.2	1,820
Molasses		....	....	24.6	75.4	....	....	73.1	2.3	1,360

TABLE 3.—*Composition of water-free substance of edible portion of side of beef of medium fatness.*

Portion taken for analysis.	Nitrogen.	Protein (N. × 6.25).	Fat.	Ash.	Protein, fat, and ash.	Protein by difference.
	Per cent.	*Per cent.*	*Per cent.*	*Per cent.*	*Per cent.*	*Per cent.*
First cut neck	7.64	47.75	51.19	2.42	101.36	46.39
Second cut neck	9.09	56.83	40.80	3.02	100.65	56.18
Third cut neck	8.19	51.19	45.49	2.69	99.37	51.82
First cut chuck ribs	5.99	37.44	61.34	2.09	100.87	36.57
Second cut chuck ribs	7.46	46.63	48.48	2.79	97.90	48.73
Third cut chuck ribs	8.37	52.31	45.90	3.02	101.23	51.08
First cut ribs	4.58	28.63	69	1.64	99.27	29.36
Second cut ribs	4.63	28.94	69.85	1.67	100.46	28.48
Third cut ribs	4.82	30.13	68.86	1.70	100.69	29.44
Brisket	4.16	26	70.69	1.54	98.23	27.77
Shoulder clod	9.94	62.13	34.55	3.56	100.24	61.89
Cross ribs	3.94	24.63	74.13	1.36	100.12	24.51
Shin	13.99	87.46	8.78	4.76	101	86.46
Plate	4.35	27.19	70.85	1.44	99.48	27.71
Navel	4.75	29.69	69.52	1.58	100.79	28.90
Small end sirloin	6.72	42	54.80	2.22	99.02	42.98
Hip sirloin	6.68	41.76	55.76	2.29	99.81	41.95
Socket	6.29	39.32	58.76	2.37	100.45	38.87
Rump	4.02	25.13	74.18	1.31	100.62	24.51
First cut round	9.20	57.50	39.48	3.17	100.15	57.35
Second cut round	10.87	67.94	28.16	4.36	100.46	67.48
Leg	11.99	74.94	20.23	4.31	99.58	75.36
Top of sirloin	3.80	23.75	75.57	1.31	100.63	23.12
Flank	2.75	17.19	82.55	.90	100.64	16.55
Kidney fat	.17	1.06	98.88	.20	100.14	.92

TABLE 4.—*Composition of flesh (edible portion) of side of beef of medium fatness.*

Portion taken for analysis.	Water.	Water-free substance.	Protein by difference.	Fat.	Ash.	Protein (N. × 6.25).	Water, protein, fat, and ash.
	Per cent.	*Per cent.*	*Per cent.*	*Per cent.*	*Per cent.*	*Per cent.*	*Per cent.*
First cut neck	60.64	39.36	18.20	20.15	0.95	18.79	100.53
Second cut neck	64.48	35.52	19.96	14.49	1.07	20.18	100.22
Third cut neck	61	39	20.21	17.74	1.05	19.96	99.75
Total neck	61.99	38.01	19.25	17.74	1.02	....	....
First cut chuck ribs	53.11	46.89	17.15	28.76	.98	17.56	100.41
Second cut chuck ribs	58.21	41.79	20.38	20.25	1.16	19.50	99.12
Third cut chuck ribs	63.67	36.33	18.57	16.67	1.09	18.95	100.38
Total chuck ribs	58.06	41.94	19.06	21.78	1.10	....	....
First cut ribs	46.21	53.79	15.82	37.09	.88	15.38	99.56
Second cut ribs	48.64	51.36	14.63	35.87	.86	14.86	100.23
Third cut ribs	48.48	51.52	15.16	35.48	.88	15.52	100.36
Total ribs	47.82	52.18	15.22	36.08	.88	....	....
Brisket	47.41	52.59	14.58	37.20	.81	13.68	99.10
Shoulder clod	66.61	33.39	20.66	11.54	1.19	20.74	100.08
Cross ribs	43.95	56.05	13.73	41.56	.76	13.80	100.07
Shin	73.80	26.20	22.66	2.30	1.24	22.90	100.24
Plate	44.36	55.64	15.41	39.43	.80	15.13	99.72
Navel	47.59	52.41	15.14	36.44	.83	15.56	100.42
Total fore quarter	54.12	45.88	17.27	27.65	.96	....	....
Small end sirloin	60.68	39.32	16.92	21.53	.87	16.50	99.58
Hip sirloin	58.86	41.14	17.26	22.94	.94	17.18	99.92
Small end and hip sirloin	59.86	40.14	17.07	22.17	.90	....	....
Socket	57.12	42.88	16.67	25.19	1.02	16.86	100.19
Rump	40.23	59.77	14.65	44.34	.78	15.02	100.37
First cut round	66.04	33.96	19.48	13.40	1.08	19.53	100.05
Second cut round	69.53	30.47	20.57	8.57	1.33	20.71	100.14
First cut and second cut round	66.76	33.24	19.71	12.40	1.13	....	....
Leg	72.15	27.85	20.99	5.66	1.20	20.87	99.88
Top of sirloin	42.20	57.80	13.36	43.68	.76	13.75	100.39
Flank	27.45	72.55	12.01	59.89	.65	12.47	100.46
Total hind quarter, except kidney fat	55.66	44.34	17.11	26.27	.96	....	....
Kidney fat	4.30	95.70	.89	94.62	.19	1.02	100.13
Whole side	52.43	47.57	16.44	30.20	.93	....	....
Whole side, except kidney fat	54.77	45.23	17.20	27.07	.96	....	....

TABLE 5.—*Composition of side of beef of medium fatness as received, including both edible portion and refuse.*

Portion taken for analysis.	Refuse.	Edible portion.	Edible portion.				
			Water.	Water-free substance.	Nutrients.		
					Protein by difference.	Fat.	Ash.
	Per cent.	Per cent.	Per cent.	Per cent.	Per cent.	Per cent.	Per cent.
First cut neck	18.37	81.63	49.50	32.13	14.90	16.45	0.78
Second cut neck	18.74	81.26	52.40	28.86	16.22	11.77	.87
Third cut neck	22.86	77.14	47.06	30.08	15.59	13.68	.81
Total neck	19.54	80.46	49.88	30.58	15.49	14.27	.82
First cut chuck ribs	19.45	80.55	42.78	37.77	13.82	23.16	.79
Second cut chuck ribs	19.59	80.41	46.81	33.60	16.39	16.28	.93
Third cut chuck ribs	16.31	83.69	53.29	30.40	15.54	13.95	.91
Total chuck ribs	18.80	81.20	47.14	34.06	15.48	17.69	.89
First cut ribs	21.36	78.64	36.34	42.39	12.44	29.17	.69
Second cut ribs	22.70	77.30	37.60	39.70	11.31	27.73	.66
Third cut ribs	20.35	79.65	38.61	41.04	12.07	28.26	.71
Total ribs	21.34	78.66	37.62	41.04	11.97	28.38	.69
Brisket	14.33	85.67	40.62	45.05	12.49	31.87	.69
Shoulder clod	15.62	84.38	56.21	28.17	17.43	9.74	1
Cross ribs	12.17	87.83	38.60	49.23	12.06	36.50	.67
Shin	40.16	59.84	44.16	15.68	13.56	1.38	.74
Plate	17.90	82.10	36.42	45.68	12.65	32.13	.66
Navel	11.41	88.59	42.16	46.43	13.41	32.28	.74
Total fore quarter	18.45	81.55	44.13	37.42	14.09	22.55	.78
Small end sirloin	24.46	75.54	45.84	29.70	12.78	16.26	.66
Hip sirloin	27.32	72.68	42.78	29.90	12.54	16.68	.68
Small end and hip sirloin	25.79	74.21	44.42	29.79	12.67	16.45	.67
Socket	35.79	64.21	36.68	27.53	10.70	16.17	.66
Rump	16.18	83.82	33.72	50.10	12.28	37.17	.65
First cut round	7.74	92.26	60.93	31.33	17.97	12.36	1
Second cut round	32.12	67.88	47.20	20.68	13.96	5.82	.90
First cut and second cut round	14.13	85.87	57.33	28.54	16.92	10.65	.97
Leg	62.22	37.78	27.26	10.52	7.93	2.14	.45
Top of sirloin	3.23	96.77	40.84	55.93	12.93	42.27	.73
Flank	11.47	88.53	24.30	64.23	10.63	53.02	.58
Total hind quarter, except kidney fat	20.23	79.77	44.40	35.37	13.65	20.96	.76
Kidney fat	....	100	4.39	95.70	.89	94.62	.19
Whole side	18.48	81.52	42.74	38.78	13.40	24.62	.76
Whole side, except kidney fat	19.21	80.79	44.25	36.54	13.90	21.87	.77

TABLE 6.—*Composition of side of beef of medium fatness (weights of ingredients in meat as received).*

Portion taken for analysis.	In whole specimen, as taken for analysis.						
				Edible portion.			
	Total weight of cut.	Refuse.	Edible portion.	Water.	Protein by difference.	Fat.	Ash.
	Pounds.	*Pounds.*	*Pounds.*	*Pounds.*	*Pounds.*	*Pounds.*	*Pounds.*
First cut neck	8.09	1.48	6.61	4.01	1.21	1.33	0.06
Second cut neck	6.05	1.13	4.92	3.18	.98	.71	.05
Third cut neck	4.33	.99	3.34	2.04	.67	.59	.04
Total neck	18.47	3.60	14.87	9.23	2.86	2.63	.15
First cut chuck ribs	11.75	2.28	9.47	5.03	1.63	2.72	.09
Second cut chuck ribs	20.60	4.04	16.56	9.64	3.38	3.35	.19
Third cut chuck ribs	9.44	1.54	7.90	5.03	1.47	1.32	.08
Total chuck ribs	41.79	7.86	33.93	19.70	6.48	7.39	.36
First cut ribs	7.23	1.54	5.69	2.63	.90	2.11	.05
Second cut ribs	6.82	1.55	5.27	2.56	.77	1.89	.05
Third cut ribs	9.41	1.91	7.50	3.63	1.14	2.66	.07
Total ribs	23.46	5	18.46	8.82	2.81	6.66	.17
Brisket	13.64	1.96	11.68	5.54	1.70	4.35	.09
Shoulder clod	17.15	2.68	14.47	9.64	2.99	1.67	.17
Cross ribs	14.70	1.80	12.96	5.69	1.78	5.39	.10
Shin	11.11	4.46	6.65	4.91	1.51	.15	.08
Plate	15.37	2.75	12.62	5.60	1.94	4.98	.10
Navel	19.51	2.23	17.28	8.22	2.62	6.30	.14
Total fore quarter	175.26	32.34	142.92	77.35	24.69	39.52	1.36
Small end sirloin	13.41	3.29	10.15	6.16	1.72	2.18	.09
Hip sirloin	11.74	3.21	8.53	5.02	1.47	1.96	.08
Small end and hip sirloin	25.16	6.50	18.68	11.18	3.19	4.14	.17
Socket	9.66	3.46	6.20	3.55	1.03	1.56	.06
Rump	12.61	2.04	10.57	4.25	1.55	4.69	.08
First cut round	37.52	2.90	34.62	22.87	6.74	4.64	.37
Second cut round	13.33	4.28	9.05	6.29	1.86	.78	.12
First cut and second cut round	50.85	7.18	43.67	29.16	8.60	5.42	.49
Leg	8.58	5.34	3.24	2.34	.68	.18	.04
Top of sirloin	10.24	.33	9.91	4.18	1.33	4.33	.07
Flank	13.18	1.51	11.67	3.21	1.40	6.99	.07
Total hind quarter, except kidney fat	130.30	26.36	103.94	57.87	17.78	27.31	.98
Kidney fat	12	....	12	.52	.11	11.35	.02
Whole side	317.56	58.70	258.86	135.74	42.58	78.18	2.36
Whole side, except kidney fat	305.56	58.70	246.86	135.22	42.47	66.83	2.34

TABLE 7.—*Composition of water-free substance of edible portion of side of mutton and side of lamb.*

Portion taken for analysis.	Nitrogen.	Protein (N. × 6.25).	Fat.	Ash.	Protein, fat, and ash.	Protein by difference.
Side of mutton:	*Per cent.*	*Per cent.*	*Per cent.*	*Per cent.*	*Per cent.*	*Per cent.*
Shoulder	6.97	43.56	54.09	2.28	99.93	43.63
Breast	3.51	21.94	75.60	1.62	99.16	22.78
Neck	5.75	35.94	61.76	1.79	99.49	36.45
Rack	6.47	40.44	57.50	1.72	99.66	40.78
Leg	7.59	47.43	49.69	2.36	99.48	47.95
Loin	4.69	29.31	69.01	1.43	99.75	29.56
Flank	4.07	25.44	73.48	.75	99.67	25.77
Kidney and kidney fat	1.21	7.56	94.20	.50	102.26	5.30
Side of lamb:						
Shoulder	6	37.50	61.54	2.12	101.16	36.34
Breast	6.96	43.50	53.91	2.31	99.72	43.78
Neck	6.53	40.81	57.28	2.27	100.36	40.45
Leg	8.95	55.94	43.38	3.04	102.36	53.58
Loin	6.89	43.06	55.60	2.34	101	42.06

TABLE 8.—*Composition of flesh (edible portion) of side of mutton and of side of lamb.*

Portion taken for analysis.	Water.	Water free substance.	Protein by difference.	Fat.	Ash.	Protein (N. × 6.25).	Water, protein, fat, and ash.
Side of mutton:	*Per cent.*	*Per cent.*	*Per cent.*	*Per cent.*	*Per cent.*	*Per cent.*	*Per cent.*
Shoulder	58.56	41.44	18.08	22.41	0.95	18.05	99.97
Breast	37.60	62.40	14.22	47.17	1.01	13.69	99.47
Neck	55.69	44.31	16.16	27.36	.79	15.93	99.77
Rack	54.94	45.06	18.38	25.91	.77	18.22	99.84
Fore quarter	52.22	47.78	16.96	29.88	.94	....	....
Leg	61.80	38.20	18.32	18.98	.90	18.12	99.80
Loin	49.27	50.73	14.99	35.01	.73	14.87	99.88
Flank	38.73	61.27	15.78	45.03	.46	15.59	99.81
Hind quarter	54.67	45.33	16.92	27.53	.88	....	....
Kidney and kidney fat	18.81	81.19	4.30	76.48	.41	6.14	101.84
Side of lamb:							
Shoulder	51.83	48.17	17.51	29.64	1.02	18.06	100.55
Breast	56.24	43.76	19.16	23.59	1.01	19.04	99.88
Neck	56.69	43.31	17.54	24.81	.96	17.67	100.13
Fore quarter	55.06	44.94	18.12	25.82	1	....	....
Leg	64.72	35.28	18.91	15.30	1.07	19.74	100.83
Loin	54.82	45.18	19	25.12	1.06	19.45	100.45
Hind quarter	60.88	39.12	18.96	19.09	1.07	....	....
Whole side, except kidney and kidney fat	57.94	42.06	18.53	22.50	1.03	....	....

TABLE 9.—*Composition of side of mutton and side of lamb as received, including both edible portion and refuse.*

Portion taken for analysis.	Refuse.	Edible portion.	Edible portion.				
			Water.	Water-free substance.	Nutrients.		
					Protein by difference.	Fat.	Ash.
Side of mutton:	*Per cent.*	*Per cent.*	*Per cent.*	*Per cent.*	*Per cent.*	*Per cent.*	*Per cent.*
Shoulder	16.29	83.71	49.02	34.69	15.14	18.76	0.79
Breast	14.95	85.05	31.98	53.07	12.09	40.12	.86
Neck	27.58	72.42	40.33	32.09	11.70	19.82	.57
Rack	19.28	80.72	44.35	36.37	14.84	20.91	.62
Fore quarter	18.97	81.03	42.31	38.72	13.74	24.22	.76
Leg	18.12	81.88	50.60	31.28	15	15.54	.74
Loin	15.75	84.25	41.51	42.74	12.63	29.50	.61
Flank	2.15	97.85	37.90	59.95	15.44	44.06	.45
Hind quarter	15.65	84.35	46.11	38.24	14.27	23.23	.74
Kidney and kidney fat	....	100	18.81	81.19	4.30	76.48	.41
Whole side, except kidney fat	17.30	82.70	44.23	38.47	14.01	23.71	.75
Side of lamb:							
Shoulder	20.33	79.67	41.29	38.38	13.96	23.61	.81
Breast	19.09	80.91	45.50	35.41	15.50	19.09	.82
Neck	17.67	82.33	46.67	35.66	14.44	20.43	.79
Fore quarter	18.84	81.16	44.69	36.47	14.71	20.95	.81
Leg	17.70	82.30	53.26	29.04	15.56	12.60	.88
Loin	12.18	87.82	48.14	39.68	16.69	22.06	.93
Hind quarter	15.65	84.35	51.35	33	15.99	16.11	.90
Whole side, except kidney fat	17.30	82.70	47.92	34.78	15.32	18.61	.85

TABLE 10.—*Composition of side of mutton and side of lamb (weights of ingredients in materials as received).*

Portion taken for analysis.	Total weight of cut.	Refuse.	Edible portion.	Edible portion.			
				Water.	Nutrients.		
					Protein by difference.	Fat.	Ash.
Side of mutton:	*Pounds.*	*Pounds.*	*Pounds.*	*Pounds.*	*Pounds.*	*Pounds.*	*Pounds.*
Shoulder	3.17	0.52	2.65	1.55	0.48	0.59	.03
Breast	1.98	.29	1.69	.63	.24	.80	.02
Neck	1.73	.48	1.25	.69	.20	.35	.01
Rack	2.29	.45	1.84	1.01	.34	.48	.01
Fore quarter	9.17	1.74	7.43	3.88	1.26	2.22	.07
Leg	5.20	.94	4.26	2.62	.79	.81	.04
Loin	3.26	.51	2.75	1.30	.41	.96	.02
Flank	.93	.02	.91	.35	.14	.41	.01
Hind quarter	9.39	1.47	7.92	4.33	1.34	2.18	.07
Kidney and kidney fat	.39	....	.39	.07	.02	.30	....
Whole side	18.95	3.21	15.74	8.28	2.62	4.70	.14
Whole side, except kidney fat	18.56	3.21	15.35	8.21	2.60	4.40	.14
Side of lamb:							
Shoulder	2.85	.57	2.28	1.18	.40	.68	.02
Breast	3.57	.68	2.89	1.63	.55	.68	.03
Neck	3.03	.53	2.50	1.42	.44	.62	.02
Fore quarter	9.45	1.78	7.67	4.23	1.39	1.98	.07
Leg	5.50	.97	4.53	2.93	.86	.69	.05
Loin	3.32	.41	2.91	1.60	.55	.73	.03
Hind quarter	8.82	1.38	7.44	4.53	1.41	1.42	.08
Whole side, except kidney fat	18.27	3.16	15.11	8.70	2.80	3.40	.15

WORK NOW NEEDED IN ANALYSIS OF FOODS.

In the present condition of our knowledge of the composition of materials used for the food of man and with the results which have accumulated up to the present, investigation is especially needed in two directions: (1) The study of the methods of analysis with a view to their improvement, and (2) analyses of a sufficient number of specimens to give a clear idea of the range in composition and the average proportions of ingredients in the materials in common use in the United States. The study of methods is one of the pressing necessities of physiological chemistry at the present time. So many analyses of food materials have been made by the current methods that it is hardly desirable to devote a very large amount of labor to further analyses except for specific purposes, such as the study of dietaries, i. e., the actual food consumption of people in different places and under different conditions of life, and the study of the food supply of particular places and of the composition of certain classes of food materials of which but few analyses have thus far been made. If the studies of dietaries should be carried out in different parts of the country in the manner and to the extent which now seem to be desirable the analyses involved will bring the larger part of the information that is most needed regarding the composition of our ordinary food materials.

IMPROVEMENT OF METHODS OF ANALYSIS.

Among the things of fundamental importance for furthering the knowledge of the value and proper uses of foods is the improvement of methods of analysis. This necessity will be clearer if we consider what are the things we have to analyze and exactly why we analyze them. We have to distinguish here between the materials which are used as the food of man and those which are fed to domestic animals. The former may be conveniently designated as food materials or foods and the latter as fodder materials or feeding stuffs.

The methods of analysis used for foods are similar to those for feeding stuffs, and the two classes of materials need to be studied together.

Classification of ingredients.—We make analyses of foods and feeding stuffs to learn their value for nutriment and the proper ways to use them. In doing so we classify the ingredients in different groups, as stated on pages 11 and 12, and assign to each group a specific nutritive value.

Present usage makes the groups practically the same for all vegetable substances, thus ignoring the wide differences of kindred compounds in different plants and parts of plants. We even go so far as to make nearly the same grouping for animal as for vegetable compounds. Not only do we thus place compounds of widely different chemical and physiological characters in the same group, but we frequently put into a group compounds which do not belong there at all.

In our analyses we attempt to separate the ingredients and determine their respective amounts. We base our methods of analysis mainly upon two classes of properties, the elementary composition of the compounds, and their solubilities; and yet our knowledge of these properties is imperfect at best, and in some cases scarcely suffices for more than to assure us of the incorrectness of our methods of estimation. In many instances, especially with vegetable materials, the solubility of the compounds in laboratory reagents, their digestibility in laboratory tests by the so-called method of artificial digestion, their actual digestibility in the body of the animal, and their nutritive effect are dependent upon the ways in which they are held in the vegetable tissues, e. g., the nature of the cell walls or incrusting substances. Of these things ordinary chemical analysis tells little or nothing, and we must look to the vegetable histologist for information about them. Meanwhile, in ignorance of them we commit more or less serious error.

In vegetable materials by our present method we determine, or assume that we determine, one group, which we call protein, by multiplying the total nitrogen by 6.25. In many cases we know this factor is wrong, in many others we have no proof at all that it is right. We assume to determine a second group, which we call fats or crude fats, by extracting with ether. We know that, especially in vegetable feeding stuffs, the ether extracts a good deal of material which can not be properly classed with the fats and that it may leave more or less of the true fats unextracted. We estimate what we call fiber or crude fiber by extracting with dilute acid and alkali, and do not know the constitution of the residue, although we are certain that it varies with the conditions of extraction, and that it may contain more or less of other materials than cellulose and lignin. We determine what we call ash by incineration, and take the residue as representing the mineral matters, although the method does not tell us their forms of combination and the result does not even represent their exact amount. Even the current methods of determining water are extremely faulty. We keep the substance for a certain time at a temperature near that of boiling water, either in air or in a current of hydrogen, and reckon the loss of weight as water. We do not know how much water remains in what we call the water-free residue, but it is certain that in many cases more or less of the organic matter is volatilized and sometimes the quantity which thus escapes is very considerable. And when we have made these rough estimates of water, protein, fat, crude fiber, and ash, we add them together, subtract their sum from the whole weight, and call the difference "non-nitrogenous extract" or "carbohydrates." When this is done we make more or less rough estimates of the digestibility of the several classes of nutrients by use of digestion coefficients of doubtful accuracy. And finally, having thus estimated the total and digestible nutrients, we apply certain figures for estimated heats of combustion of protein, fats, and carbohydrates, and thus calculate the fuel values of our foods and feeding stuffs. Notwithstanding all these difficulties and sources of error it is probable that the results for animal foods are not very far from the truth. With many vegetable foods the case is no worse. Perhaps it is safe to say that the present methods of analyses of animal and vegetable foods are accurate enough to give tolerably fair estimates of their nutritive values, but they are very far from what they should be. The demand for investigation which shall lead to improvement is immediate and pressing.

With vegetable feeding stuffs the case is much worse, so bad indeed that some of the leading experimenters in the United States are inclined to make but little use of chemical analyses in feeding experiments. In my judgment this is going too far, decidedly so. But nevertheless the fact obtains that in estimates of the nutritive values of feeding stuffs, the results of our feeding trials and the teachings of experience are often at variance, and it is not easy to see why. We shall not make them harmonize until we learn how to make our analyses more correct. If the need of improvement of the methods of analysis of food is great the need of improving the methods for feeding stuffs is still greater.

At the same time there is much to encourage us. If we are still using the so-called "Weende methods" of thirty years ago with comparatively little modification we have found out how to get results that are pretty nearly uniform. To this latter consummation the work of such organizations as the Association of Official Agricultural Chemists in the United States and the Association of Experiment Stations in Germany have materially contributed; and it is no disparagement of the great value of that work to say that it is not of exactly the kind that is needed to bring the best results.

The general plan of the cooperative experiments made by these associations is that of testing certain prescribed methods of analysis. The method of analysis to be followed may be either one in common use or a new one which is to be tested. The details of the operation are minutely described and the substance or substances in which the determinations are to be made are definitely agreed upon. A single sample of a substance of a given class is selected by one analyst and distributed to the others, or each analyst selects the substance for himself. The analyses are made, reported, and compared. Generally little or no effort is made to determine the actual constituents of the compounds that are being dealt with or the conditions of combination and separation which so materially affect the accuracy of the analysis. The work is apt to be mainly a comparison of methods of manipulation in the hands of different men. This is valuable, indeed indispensable, but it does not go to the root of the matter. It is empirical and not truly scientific. It does not reveal the real sources of error and the ways of avoiding them. It is working systematically but not philosophically. It is an attempt to find a path by groping in the dark, where light is needed to show the obstacles, the pitfalls, and the safest and nearest route.

Sources of error and the investigations needed to develop correct methods.—The time has come when systematic effort ought to be made to improve our current methods of analysis. To decide what are the first steps we shall do well to consider the especial difficulties. To do this satisfactorily would require a treatise of considerable length. The present purpose will be served by very brief recapitulation of some of the main points.

Water.—It is becoming evident to experienced analysts that the accurate determination of water is one of the most difficult in this branch of analysis. Just what are the sources of error we do not fully know. We are not certain, for instance with respect to any ordinary material, just how long heating, either at a temperature of 95°, which is common in an ordinary drying oven, or at the exact temperature of boiling water which we sometimes attempt to maintain, or at the higher temperature of 105° or 110° which is occasionally recommended, is necessary for complete expulsion of the water. But we do know that at either of these temperatures there is more or less danger of volatilization of organic substances. Pure filter paper is gradually decomposed and finally charred at temperatures much below 100°. The volatile fatty matters may escape at even lower temperatures. Nitrogenous compounds are likewise driven off under the same circumstances. Some fatty matters absorb oxygen and increase in weight when dried in air; hence the necessity of drying in hydrogen. Not knowing what better to do, we have recourse to the entirely empirical method of drying a given weight of a given substance for a given time at a given temperature, and assuming that if the water is not entirely expelled enough organic matter will be driven off to compensate for the error. But it is clear that in so doing we are not getting accurate results.

Nitrogenous substances.—The method of multiplying the total nitrogen by 6.25 and taking the product as a measure of the nitrogenous substances is a most unsatisfactory makeshift. It does not tell the total quantity of nitrogenous substances, still less does it tell the amount in any given form of combination. Conventionally we call the product protein. We assume it to include in animal foods:

(1) The albuminoids; namely, native albumins, derived albumins, including casein, globulins, coagulated proteids, albumoses, and peptones.

(2) The so-called "gelatinoids," namely, those which make the basis of connective tissue, and derivatives from them. The most important of these are the ones which yield gelatin and chondrin, though the substances which yield elastin and mucin are generally classed with them.

(3) Cleavage products of albuminoids and gelatinoids other than proteoses and peptones. The most important of these for our present purpose are the so-called non-nitrogenous extractives, the group to which kreatin and kreatinin belong.

In the flesh of vertebrates, at least of those most commonly used for food, the total nitrogen multiplied by 6.25 is found to very nearly equal the total nitrogenous substances; in other words, the nitrogen factor 6.25 is a fairly accurate one. But the albuminoids, gelatinoids, and extractives have different functions in nutrition. It is especially important to devise means for accurately separating the extractives.

Considerable work on the methods of analysis of muscular tissue has been performed in the writer's laboratory, the larger part by Dr. H. B. Gibson. We have thus far found no easy and accurate way for separating the nitrogenous extractives, nor have we been able to make an at all satisfactory separation of the albuminoids and gelatinoids by the use of either hot water or dilute (one-fifth per cent) hydrochloric acid. The method which is commonly recommended for separating the gelatin by boiling water proves to be unreliable. In the muscular tissue of beef, from which the fat had been removed as completely as was convenient, and in the flesh of codfish nitrogenous material was dissolved as long as the treatment with hot water was continued, and it was evident that under the treatment some kind of decomposition was constantly going on. This was not surprising in view of the decompositions which take place in the treating of pure albuminous matter, such as blood fibrin, by very dilute acids or alkalis with heat. It accords with the observations that various albuminoids may be changed to substances allied to if not identical with peptones by such treatment. The experience with the dilute hydrochloric acid, which is supposed to dissolve the albuminoids and leave the gelatin-like compounds undissolved, was equally unsatisfactory.

A further result of the investigation just referred to, which still awaits publication, is worth noting here. Muscular tissue of beef and of codfish was partially dried by heat in air in the way commonly followed in preparing substances for further analysis. This heating in air was continued until the substance was dry and hard enough to be pulverized. The total quantity of nitrogen in the fresh substance and that in the same substance after drying were determined by the Kjeldahl method. It was found that from 1 to 3 per cent of the nitrogen was lost in the process of partial drying. The operation was repeated in a current of hydrogen, which, after passing over the substance, was conducted through hot sulphuric acid. In this way the nitrogen of the volatilized material was approximately determined. In some of the cases with meat it reached 2 per cent, and in some white fleshed fish 2½ per cent of the total nitrogen in the substance. The nitrogen thus volatilized, added to the quantity in partially dried material from which it had escaped, was found in some cases to be almost exactly equal to that in the original substance. Whether the volatilized nitrogenous compounds were nitrogenous extractives or other and more volatile cleavage products was not determined. We suspect that they might have belonged to the latter class, one reason being that the fish was hardly as fresh as the meat and that the flesh of fish is known to yield volatile amins as decomposition products.

This experience has an important bearing upon the determination of nitrogen in animal substances. It is evident that if we partially dry the substance before making the determination of nitrogen we run the risk of loss of nitrogenous material; and yet this is exactly what is done in the ordinary method of analysis. It also has an important bearing upon the determination of water. Ordinarily we estimate the loss of water in the partial drying and then take a smaller portion of the partially dry substance, heat it further until it ceases to lose weight, and call the total loss water. If the water is all expelled and as much volatile organic matter escapes with it, as the observation just mentioned would imply, the water determinations must be wrong.

The studies just referred to imply that there may be a loss of nitrogen in the complete drying which follows the partial drying in the ordinary routine of analysis of animal tissues. This would affect the final result in the same ways as the loss in partial drying.

The nitrogenous substances in vegetable foods include: (1) Albuminoids of various kinds; (2) so-called amids, i. e., synthesis and cleavage products of various kinds.

The vegetable albuminoids appear to be more variable in composition than the animal albuminoids, and the nitrogen factor for them needs to be fitted to the compounds as they actually occur in different classes of food materials. Such investigations as those of Osborne and Chittenden[1] upon this subject are valuable; indeed inquiry of this kind is of fundamental importance.

---

[1] Connecticut State Experiment Station Reports, passim; Am. Chem. Jour., vols. 13, 14, 15, passim; Jour. Am. Chem. Soc., XVI, 703.

So-called "amids" occur to only a small extent in most of our vegetable foods. In potatoes, however, the quantity is very large and their further study is more to be desired. In feeding stuffs, especially the grasses and forage plants, which are ordinarily harvested during the period of growth and marked cellular activity, the proportions are often large.

The method of Stutzer for the separating of so-called "albuminoids" and "non-albuminoids" of vegetable substances is frequently used and is regarded by some as measurably satisfactory. In the writer's laboratory the experience has not been as favorable as is to be desired, and there is little risk in saying that further investigation of the subject is much needed.

The practice of determining nitrogenous substances by multiplying the total nitrogen by 6.25 is at best a temporary makeshift. In meats, fish, milk, and the cereal grains it is probably not far out of the way, but in all of these, and especially in the meats and fish, it is of the greatest importance to distinguish between the albuminoid nitrogen and that which exists in other forms. The same is true to equal if not greater degree in potatoes and other roots, in ordinary vegetables, and in many feeding stuffs.

Terminology.—One matter that demands attention is the terminology of the nitrogenous substances. This is in almost hopeless confusion. American, English, French, German, and other chemists disagree widely and most curiously in their usage. By a large number the term albuminoid is used in the sense in which it is used here. But it is employed by others not only for other allied compounds, but even for such substances as hæmoglobin. The word "proteids" also meets promiscuous handling by different authorities. Some make it synonymous with albuminoids; with others it includes what are here called "albuminoids" and "gelatinoids;" by others it is applied to a different and greater variety of nitrogenous materials. And while the term is applied by most physiological chemists, so far as I know, to specific compounds, it is occasionally used by writers on the subject for the total nitrogenous substances in food materials without regard to their composition.

It is, of course, desirable that a nomenclature for the nitrogenous compounds be agreed upon. Provisionally it seems to the writer wise to conform to what is coming to be a very common usage with respect to one term, "protein;" namely, to apply this to the total nitrogenous substance (exclusive of nitrogenous fats) whether obtained by multiplying the nitrogen by 6.25, which is now commonly accepted as the measure of the nitrogenous substance, or by difference as is done with the animal tissues which contain very little carbohydrates.

Fats—Ether extract.—The case with the so-called "crude fat" is no better than with the nitrogenous compounds. In such animal substances as muscular or connective tissue, and milk and its products, when they are fresh, i. e., when the cleavage of proteids which comes with exposure and the action of bacteria has not proceeded too far, the ether extract, if obtained by proper manipulation, contains probably all the so-called "neutral" fats or "triglycerids" (i. e., the normal propenyl compounds with the fatty and oleic acids), and little else.

The amounts of lecithins in these tissues are very small and very probably their extraction by ether may be complete enough for practical purposes.

In eggs the extraction of lecithin with ether alone is hardly feasible by the ordinary method, indeed, the experience in the writer's laboratory has shown it to be extremely slow and difficult with ether or alcohol, or even with a mixture of ether and alcohol, although I do not feel competent to say that it can not be done easily and quickly by proper manipulation. This is one of the minor details of method that need further investigation.

The compounds commonly extracted from vegetable products and denominated "crude fat," "ether extract," or simply "fats," are very diverse; nor are we at all sure what proportion of them we extract by our ordinary methods. We have to deal not only with the true fats, i. e., glycerids of the fatty acids and the fatty acids themselves, which may be properly classed with the fats in estimating nutritive values, but also with a great variety of other compounds of widely differing constitution, and of whose functions and value in nutrition but little is known. Among them are substituted glycerids, including lecithins, waxes, alkaloids, cholesterin, hydrocarbons, and chlorophylls. Of these, the lecithins appear to have a special value in nutrition, while some of the alkaloids are poisons. An especial and serious difficulty is found with the fats which become insoluble in ether when heated in air. These seem to be abundant in some seeds and grasses. We assume that they are of the nature of the drying oils, like linolein, and take up oxygen on exposure to air. We therefore dry the substance in a current of hydrogen before extracting with ether. This is very essential. The writer has found the quantity of material dissolved by ether from seeds, e. g., of maize, to be reduced three-fourths by drying in air. We have not found this difficulty to occur in animal tissues, and probably the general experience is to the same effect. Unfortunately there is still much doubt as to just what are the bodies which become insoluble in ether by drying in air or the list of materials in which they occur. It is doubtless wise to use hydrogen, especially in the analyses of vegetable materials and to make sure that the hydrogen is free from oxygen. The use of illuminating gas instead of hydrogen for this purpose, which has been found satisfactory in some laboratories,[1] is worthy of more consideration.

[1] See Foerster, Landw. Vers. Stat., 37, 1890, 57 and Experiment Station Record, 5, p. 383; and Märcker, Bieler and Schneidewind, Die agrikultur-chemische Versuchsstation Halle a/S., p. 11.

If the substance to be analyzed is finely ground, free from water, and in otherwise normal condition, and the cell structure allows easy extraction we may in general expect that ether, applied in accordance with the official method, which is commonly followed by our stations, will take out the whole of the true fats and fatty acids, and more or less of the lecithins, wax, chlorophyll, cholesterin, alkaloids, and hydrocarbons, and that more or less of the compounds other than true fats and fatty acids will remain undissolved. If the ether contains alcohol the extract may be expected to contain more of the other compounds. If water is present the extract may be larger, as has actually been found to be the case in numerous observations. In other words, the extract, as obtained by the ordinary method, contains either part or all of the fats and fatty acids, as the case may be, and with them more or less of other substances.

With animal foods, except eggs and such materials as the edible portion of oysters in which different organs are included, it is doubtless safe to say that the ordinary method of ether extraction is reasonably satisfactory.

With vegetable foods the case is complicated by the presence of other compounds than the true fats (normal propenyl compounds) and by the fats that become insoluble in drying. But with the cereal products and potatoes, which make up the bulk of our vegetable foods, the ordinary method appears to do very well.

The worst difficulty is with feeding stuffs. In these we have a great variety of other substances than the true fats, and considerable amounts of the fat which oxidize in air, and furthermore we have to do with the incrusting substances which materially hinder extraction. The ether extract is apt to be mixed with materials that do not belong with the fats and to contain only part of the true fat.

There is only one way to remedy this difficulty. It is to find what the substances are which vary so greatly in solubility, in what materials, and under what conditions they occur, and how to classify, separate, and determine them.

The nutritive values of the materials which are more properly grouped as fats, i. e., the glycerids of the fatty acids, and the fatty acids themselves, are pretty well understood, though further investigation of the molecular constitution of some, and of the fuel value of all, is needed. According to the present outlook it seems probable that, although the waxes and perhaps the lecithins may be classed with the fats in estimations of the nutritive values, a separate classification of some or all of the others will be necessary; and it is clear that a more definite knowledge of the chemical constitution of all the materials, other than the neutral fats and fatty acids, is indispensable to any correct estimate of their values for nutriment.

Crude fiber—Cellulose.—The term "crude fiber," like protein, is practically a conventional expression for bridging over the present ignorance of the actual substances to which it applies, and the conventional method for its determination is another of the provisional makeshifts of the analytical laboratory.

In the analysis of food this is of minor consequence; the vegetable foods have but little cellulose and less of the incrusting substances which are associated with it, while animal foods have none. In many feeding stuffs the cellulose and ligneous substances which we lump together as crude fiber are a disturbing factor, and a very serious one, in the analysis and in the determination of the digestibility and the nutritive value.

Cellulose is a comparatively well-defined chemical compound, or to speak more accurately, the celluloses found in different plants consist of more or less clearly defined chemical compounds. The celluloses are more or less digestible even by man; they can be converted into sugar, and though their oxidation in the body in such way as to make their potential energy largely available has been called in question, later experimental inquiry ascribes to them, in so far as they are digested, practically the same nutritive value as other carbohydrates.

Intimately associated with cellulose are the ligneous substances. These are as yet but little understood. Such terms as "lignin," "zylogen," "cuticle," and "woody fiber" have been applied to them and to the hardened cell walls or other parts of plant tissue in which they occur. They abound in the more or less completely matured stems of the grasses and cereals and in the outer coatings of seeds. Hence we meet them in hay, straw, cornstalks, bran, and oil cake. They seem to be especially characteristic of what are called coarse fodders. It is these substances, indeed, with others more or less closely allied, which make such fodders indigestible and "coarse."[1] There is extremely little of them in flour and meal of cereal grains, in the seeds of the legumes, tubers, roots, or ordinary fruits. Hence they are not responsible for any large part of the difficulty in the analysis of food materials, but do make a great deal of the trouble we have with feeding stuffs, both in their analysis and in the estimates of feeding value.

Practically what we do in the so-called determination of crude fiber is to treat vegetable substances in an entirely empirical way with dilute acid and alkali, assuming that these remove the nitrogenous substances, and call the undissolved residue crude fiber. How much of cellulose or ligneous compounds may be dissolved we do not know, but we do know very well that more or less nitrogenous and other compounds are often left in the residue thus crudely separated and crudely designated.

[1] See articles on this general subject by Dr. J. B. Lindsey, entitled "The Composition of Wood," in Agricultural Science, 1893, Nos. 1-4. These include accounts of investigations by the author and a summary of the late researches upon the celluloses, lignin, and the carbohydrates, and other references to the literature of the subject.

These incrusting substances affect the whole work of analysis of vegetable foods and feeding stuffs.

As regards the bodies which we are wont to classify as "amids," "ether extract," "nitrogen-free extract," and "crude fiber," and which we attempt to estimate by treatment with water, ether, dilute acid and alkali, or other solvents, we are coming to appreciate that their solubility is influenced not only by the fineness of grinding of the sample, the time and temperature of the extraction, the quantity of the water, the purity of the ether, the strength of the acid or alkali, the time and temperature of extraction, and by chemical changes induced in the compounds by fermentation, or by long standing, or by drying in the air, but also by the ways in which they are held within the cells or occur as constituents of the plants.

The observations of the greater digestibility of cellulose in young than in older plants and in plants grown on rich as compared with those grown on poor soil, illustrate this point.

We are beginning to realize that the permeability of the cell walls and other mechanical conditions affect the ease or difficulty of extraction; that the histological structure of the plant is a most important factor; in other words, that here is one of the places where the chemist must have the help of the vegetable physiologist if he will learn to do his work as it ought to be done.

Nonnitrogenous extractives—Carbohydrates.—The use of the term "nonnitrogenous extractives" is another makeshift for covering our ignorance of the compounds we are dealing with and the imperfections of our methods of analysis. The carbohydrates as chemical compounds are coming to be well understood, thanks to the labors of such investigators as Fischer and Kiliani, which are revealing their molecular constitution, and of Tollens and F. Schulze and their associates, who are finding what they are and where they occur. These remarks apply not only to the amyloses, glucoses, and sucroses, but also to the pentoses and the gummy substances from which they are derived, although the nature of the latter is only beginning to be understood. The late progress of research in this direction is most gratifying, not only in the results achieved and promised, but also in the examples it gives of the usefulness of such abstract inquiry. It is a matter of especial congratulation that a number of gentlemen connected with colleges and experiment stations and other institutions in the United States, including Messrs. Allen, De Chalmot, Flint, Lindsey, Stone, and Wheeler, have been engaged in this field of inquiry, and it is much to be hoped that their work may be continued.

The current practice of lumping the known and unknown materials together indiscriminately and simply estimating their total amount, as is done with the nonnitrogenous extractives, is most unsatisfactory. It leaves us in the dark as to the character and nutritive values of the compounds, and is only a rough approximation to their quantity.

With the cereal grains and the flour and meals prepared from them, and with the leguminous seeds commonly used for food, the present state of affairs is not very serious. The error in water determination in drying in hydrogen is probably quite small; the nitrogenous compounds are mostly albuminoids with a nitrogen factor not very far from 6.25, though this factor evidently needs to be fitted to each kind of seed in both the cereals and the legumes; the ether extract may be expected to contain all the true fats, the fatty acids, the bulk of the lecithin, and very little else; the quantity of crude fiber is very small and consists mostly of cellulose; the content of mineral matters is so slight that the error in their determination may be disregarded; the estimate of carbohydrates by difference is thus not very far out of the way; and finally the carbohydrates themselves are reasonably well understood and their heats of combustion are readily determined.

In potatoes the nonnitrogenous extractives consist chiefly of starch, and their estimation by difference may not be very far from correct, provided a way is found for correct determination of the nitrogenous compounds.

Roots, as beets and turnips, and fruits contain large proportions of the bodies to which such names as pectose and pectin are given. Very little is known of the constitution of these compounds. Very likely they may approach the carbohydrates in nutritive value.

It is in the feeding stuffs that the nonnitrogenous extractives are most puzzling. Indeed in the coarse fodders we know extremely little of their nature, and the errors in determining the other ingredients make the estimating of their amounts by difference little better than guesswork.

We have been accustomed to think of the carbohydrates as offering no difficulty in the analyses of animal foods, and are only lately coming to see our mistake. In muscular tissue, e. g., meats and the flesh of fish, they have been entirely ignored; but the late work of Kulz, Pflüger, and others shows that the glycogen content of these substances is much too large to be ignored in accurate investigations of either their composition or their uses in nutrition.

The only carbohydrate supposed to occur to any extent in milk is lactose. Considering that the accurate determination of the water in milk is by no means one of the easy operations of the laboratory, and that any error in this, unless compensated by the slight but very probable errors in determination of protein and fats, will affect the estimate of sugar by difference, it must be allowed that even here the present method is not perfect.

Mineral matters.—That the residue left on incineration and commonly called ash does not exactly represent the mineral matters in the substance analyzed goes without saying. But so long as we are not concerned with the functions of the mineral matters this slight analytical error need not be taken into account.

Summary.—When we summarize the sources of error in our present methods of analysis it is clear that the worst difficulties are those with the feeding stuffs. Nevertheless there is serious need of reform in the methods for analysis of foods.

With animal foods the subjects of which investigation is most pressingly needed include—

(1) Determination of water.

(2) Separation of nitrogenous extractives from proteids of muscular tissue and determination of both.

(3) Study of the constitution of the proteids of animal tissues.

(4) Separation and determination of glycogen in muscular tissue.

In vegetable foods it is especially desirable to investigate—

(1) The albuminoids.

(2) The nonalbuminoid nitrogenous compounds, especially in tubers (potatoes).

(3) The lecithins and other compounds containing nitrogen and phosphorus.

With respect to all these substances the heats of combustion constitute a most important branch of the needed inquiry.

The need of an understanding of the molecular constitution of compounds in order to devise correct methods for learning their digestibility and nutritive values is illustrated by the albuminoids. What is the residue left undissolved by pepsin and trypsin to which the term "nuclein" has been applied? Is there reason for a separate classification of nucleo-albumens? Is there in these or other albuminoids a molecular group containing phosphorus which resists the digestive ferments and is the basis of the undigested portion of the compounds in which it occurs? If so, what is its relation to the nucleus or the nucleolus of the cell? Shall we not have to look to a union of organic chemistry and vegetable physiology for the facts we must have in order to devise plans for correct analyses and determinations of digestibility, potential energy, and nutritive value of the compounds?

The results of future research will doubtless lead not only to changes in the general groupings and methods of analysis, but also to special groupings and methods of analysis for different classes of vegetable and animal foods and feeding stuffs. It is hardly to be expected, for instance, that we shall always hold to the same grouping of compounds for grasses, cereal grains, leguminous plants and their seeds, root crops, milk, and meats. It is more probable that groupings for different classes of materials which shall correspond with methods of analysis and of estimating the nutritive values will prove both necessary and feasible.

It is safe to say that all of the work which has been done in the past and is being done to-day in the analysis of feeding stuffs and the feeding trials based upon them will have to be revised and much of it discarded. A large amount of experimental inquiry is being done which is not bringing the needed results and can not in the nature of the case be of the highest and most enduring value, and much of it may have to be done over again when correct methods shall have been devised.

The first step toward reform must be research in analytical, organic, physical, and physiological chemistry. The needed improvement of our methods will evidently come only as fast as does the chemical and physiological knowledge which must serve as a basis for changes. This means that the most abstract and profound study is necessary. Fortunately such study is more and more engaging the attention of chemists and vegetable physiologists.

From the chemical standpoint we need: First, such studies as will bring definite knowledge of the kinds and amounts of proximate compounds contained in each substance to be analyzed—that is to say, (1) in different species of plants, as grasses, grains, cereals, legumes, tubers, roots, etc.; (2) in plants of the same species grown under different conditions; (3) in different parts of the same plant, as the stalk and seed of maize and wheat, and the different parts of the wheat grain; (4) in the same plant at different periods of growth; (5) in animal tissues and fluids. For some of the information needed the aid of the histologist must be sought. Second, studies of each compound regarding its behavior with reagents—i. e., solubility, etc.; its elementary composition; its cleavage products; its molecular constitution; the changes it undergoes by the action of ferments; its digestibility, and its potential energy. Third, classifications of the compounds based upon the properties named. Fourth, improved methods for separation and estimation based upon the same properties. Fifth, as the outcome of all this, more correct methods of estimating the nutritive values.

Investigations in these lines have already been undertaken by the Division of Chemistry of the United States Department of Agriculture, by several of the experiment stations, and by other institutions of research. The work of the Association of Official Agricultural Chemists in developing and improving the methods of analysis has been of decided value. It has, however, confined its attention too closely to empirical comparisons of methods of manipulation.

For the collating of the results of previous inquiries, and for the prosecution of the necessary investigations, cooperation of a large number of specialists will, of course, be requisite. We may confidently expect that experiment stations will be able to devote more and more labor to these higher inquiries. The increased resources of our agricultural colleges will enable them to encourage such researches. The scientific value of this work is such that chemists in other colleges and universities ought to be led to join in it. And is it too much to suggest that international cooperation might be secured? The expense of this research may be best met by the wise expenditure of relatively small sums of money judiciously expended, so as to stimulate investigations and bring them to completion. In what the Smithsonian Institution has done in times past in promoting research by small amounts of money we have an illustration of what might be accomplished here. The result would be useful in several ways. It would encourage research, develop talent, and improve the intellectual tone of the institutions where such work was being done. Its influence upon the development of science in this country would be excellent and the practical value of the outcome would many times exceed the cost.

CHAPTER IV.

THE DIGESTIBILITY OF FOOD.

The value of food for nutriment depends not only upon how much of nutrients it contains, but also upon how much of these the body can digest and use for its support.

The question of the digestibility of foods is very complex, and it is noticeable that the men who know most about the subject are generally the least ready to make definite and sweeping statements concerning it. One of the most celebrated physiologists of the time, an investigator in whose laboratory this particular subject has been studied more than in almost any other, says in his lectures that, aside from the chemistry of the process, and the quantities of nutrients that may be digested from different foods, he is unable to affirm much about it. The contrast between this and the positiveness with which many persons discourse about the digestibility of this or that kind of food, is marked, and has its moral.

One source of confusion is the fact that, what people commonly call the digestibility of food includes several very different things; some of which, as the ease with which a given food material is digested, the time required for the process, the influence of different substances and conditions upon digestion, and the effects upon comfort and health, are so dependent upon individual peculiarities of different persons, and so difficult of measurement, as to make the laying down of hard and fast rules impossible. Why it is, for instance, that some persons are made seriously ill by so wholesome a material as milk, and others find that certain kinds of meat, of vegetables, or of sweetmeats, "do not agree with them," neither chemists nor physiologists can exactly tell. Late investigations, however, suggest the possibility that the ferments in the digestive canal may, with some people, cause particular compounds to be changed into injurious and even poisonous forms, so that it may sometimes be literally true that "one man's meat is another man's poison."[1]

But digestion proper, by which we understand the changes which the food undergoes in the digestive canal in order to fit the digestible portion to be taken into the blood and lymph and do its work as nutriment, is essentially a chemical process. About this a great deal has been learned within a comparatively few years, so that here again we have many important facts that have not yet got into current literature.[2]

---

[1] See interesting statements upon this subject in Dr. Burdon-Sanderson's "Disorders of Digestion." See also article on "Digestibility of Food" in Century Magazine, September, 1887.

[2] An excellent treatment of the general subject of the chemistry of digestion may be found in Gamgee's Physiological Chemistry, Vol. II.

Professor Maly very aptly compares food to ore, and the nutriment digested from it to the metal extracted from ore. In the chemical laboratory a metal is sometimes separated from the earthy matters with which it is mingled by pulverizing the ore, putting it in a flask, pouring acids upon it, and stirring the whole together. The acids dissolve the metal, leaving a residue of earthy matters undissolved. To separate the dissolved materials from the residue, the whole is put upon a filter, through which the solution runs, leaving the undissolved residue in the filter. Something analogous to this takes place in the digestion of food. Instead of the metal and earthy matters of the ore, we have the digestible and the undigestible constituents of meat, or bread, or other material. The grinding is done, not by pestle and mortar, but by the teeth; the digestive juices are the solvents; instead of the flask the dissolving is done in the stomach and intestine. Finally, the digested material has to pass, not through a filter, but through the porous walls of these last organs. The changes which the digestive juices cause are manifold, and not yet fully understood. But the main fact for consideration here is that the undigested residue gives a more or less accurate measuring of the digestibility of the food.

To judge accurately of the nutritive value of food, then, we must know how much of each nutrient will be digested. This is a matter that can be determined more or less accurately by experiment; but a great deal of labor is needed to make the experiments accurate. The line of research is new, the methods are not yet perfectly matured, and the results thus far obtained, though interesting and valuable when taken together, are still very far from complete. The side questions, such as differences in the digestive apparatus of different persons, the effects of exercise and rest, or the mode of preparation of the food, and of the flavoring materials and beverages taken with it, tend to complicate the problem and make satisfactory results still harder to obtain. Yet even here experimental research has brought some definite information.

THE QUANTITIES OF DIGESTIBLE NUTRIENTS IN FOOD.

The question here is, What proportion of each of the nutrients in different food materials is actually digestible? In a piece of meat, for instance, what percentages of the total protein and fats will be digested by a healthy person, and what proportion of each will escape digestion?

The proportions of food constituents digested by domestic animals has been a matter of active investigation in the European agricultural experiment stations during the past thirty years. During the past fifteen years not a little has been done in some stations in the United States. Briefly expressed, the method consists in weighing and analyzing both the food consumed and the intestinal excretion. Since the latter represents nearly the amount of food undigested, if we subtract it from the whole amount taken into the body the difference will be nearly the amount digested.

Such experiments upon human subjects, however, are rendered much more difficult by the fact that in order that the digestibility of each particular food material may be determined with certainty, it must not be mixed with other materials. Hence the diet during the experiments must be so plain and simple as to make it extremely unpalatable. An ox will live contentedly on a diet of hay for an indefinite time, but for an ordinary man to subsist a week on meat or potatoes or eggs is a very different matter. No matter how palatable such a simple food may be, at first, to a man used to the ordinary diet of a well-to-do community, it will almost certainly become repugnant to him in a few days. In consequence, the digestive functions are disturbed, and the accuracy of the trial is impaired, a fact, by the way, which strikingly illustrates the importance of varied diet in civilized life.

For instance, in one of a series of experiments conducted in the physiological laboratory at Munich by Dr. Rubner, the subject, a strong, healthy Bavarian laborer, lived for three days upon bread and water, a diet the monotony of which was much more endurable than one of meat or fish, or almost any other single-food material would have been. He was able to eat 1,185 grams of bread per day. This contained 670 grams of carbohydrates, mainly starch, of which only about 5 grams, or a little less than 1 per cent, escaped digestion. In this case, therefore, about 99 per cent of the carbohydrates of the bread was digested. The bread contained 81 grams of protein, of which 13 per cent was undigested and 87 per cent digested. The quantity of fatty matters in the bread was too small to permit an accurate test of their digestibility.

In another series, conducted by the writer in the same laboratory, the digestibility of meat in the form of beefsteak, and of fish (haddock), was tested. The subject, a medical student, consumed less than 2 pounds of meat per day, and though it was cooked with butter, pepper, salt, and onions, so as to make it to his taste, "extraordinarily well flavored," it was very difficult for him to swallow it the second day, and required still greater effort the third. The digestion, however, seemed to be normal, and all but about 1 per cent of the protein was digested.

Other trials with meat have brought similar results, and it is reasonably safe to say that when a healthy person, with sound digestive organs, eats ordinary meat or fish in proper quantity, all or nearly all of the protein is digested. Some of the fats of meat, however, seem to fail of digestion.

The number of actual experiments of this kind is very small. Nearly all have been published within the past fifteen years. The majority have been made in the physiological laboratory of the University of Munich, of which Voit is the director. Most of the subjects have been men with healthy digestive organs, two or three laboratory servants, a soldier, several medical students, and a few others. Several have been made, however, with children of a few families. All but a very small number have been conducted in Germany.

HISTORICAL SUMMARY.

Observations upon the quantities of nutrients digested from a mixed diet were made by Beneke in 1854; the earliest publication of the results that has come to our notice appeared in 1878.[1] Ranke,[2] in 1862, published accounts of quantitative tests of the digestibility of meat, but the duration of his experiments was in each case only one day, and the methods of separating the undigested residue from that of the food taken before and after the experiment had not been well elaborated at that time. In 1870 Weiske[3] published a careful study of the digestibility of crude fiber by a man with a diet consisting of cabbage, tuberous-rooted celery and carrots. Very little has since been done upon the digestibility of crude fiber by man.

A large amount of valuable work upon the digestibility of different food materials by men was carried out by Rubner[4] in the Munich physiological laboratory and published between 1879 and 1882. The subjects were a laboratory servant, who was a strong, hearty, rather stolid man of the laboring class, a soldier, and several students and assistants in the laboratory. All, or nearly all, were Bavarians.

Bread, meat, milk, eggs, and numerous other articles of food were studied and full details were published. During the past ten years inquiry in this direction has been somewhat active. Such staples as bread, meat, and milk have naturally received a large share of attention from many investigators. Besides Rubner, who has worked with all three, Prausnitz and Meyer have done much with bread; Ranke, Atwater, and Malfatti have reported results with meat; Atwater with fish; Camerer and Uffelmann with milk; and A. Mayer with butter and oleomargarine. Brief results of some Japanese work have been published but with no details as to method. Among the latest and best investigations in this line are those by Prausnitz in Voit's laboratory, in Munich.

We have found accounts of some 150 digestion experiments which seem accurate enough to be used for statistical inferences. They have been collated in tabular form and from them the selections in Table 11 have been made. Of the total number, 114 have been made with men, 5 with women, and 13 with children from 6 weeks to 14 years of age. Fourteen of the tests were made in Italy, 12 in Holland, and 3 in Japan; the rest were made in Germany. A number were made in connection with studies of dietaries; such were those in Table 11, numbered 39, 40, 41 and 42.

EXPERIMENTAL METHODS.

The real question to be considered in such a study is, What proportions of the several classes of nutrients are actually digested from

---

[1] Zur Ernährungslehre des gesunden Menschen, 1878, 299.
[2] König, Chem. d. mensch. Nahr. und Genuss., I, 1889, 37.
[3] Ztschr. Biol., 1870, 456.
[4] Ibid., 1879, 115; 1880, 119.

different food materials by persons under normal conditions as to food consumed, habits, and health? Two methods are commonly used for the study of this question, those of artificial digestion and of actual digestion experiments. The experiments on artificial digestion are made by treating the materials with solutions containing digestive ferments, as pepsin and trypsin. The most valuable work in this line has been done by Stutzer[1] and others. The experiments on actual digestion[2] are made by letting the subject, man or animal, consume a certain amount of food of known composition and comparing the quantities of water-free substance, protein, "ether extract," ash, etc., of the food and of the undigested residue, the difference being taken as the measure of the amount of the several ingredients digested.

The experiments with artificial digestion are at best only a more or less close imitation of a natural process. The value of this kind of inquiry is beyond question, but much further elaboration of details of reagents and manipulation is needed to make the method satisfactory for accurate measurement of the actual digestibility of food materials by either men or domestic animals.

The only reliable method now available for determining the digestibility of food by man is direct experiment with persons under different and well-defined conditions. Even by the use of this method we are not sure of absolutely accurate results. The chief difficulties are three, (1) the inaccuracy of the present methods of analysis of food and excreta; (2) the difficulty of distinguishing between the metabolic products and the residues of undigested food in the feces, and (3) in case of mixed diet, the impossibility of distinguishing between the undigested residues from the different materials and the effect of one food material upon the digestion of another.

The digestion of different food materials in a mixed diet has received some considerable attention recently from Prausnitz.[3] As he points out, there are three possibilities: either (1) each food digests as if it were used alone, or one is (2) hindered or (3) helped in digestion by the presence of the others. The question is a very important one. Another question, and one which demands more attention than it has received, is the effect of the methods of cooking upon the digestibility of food.[4] Very little accurate experimenting has been done in this especial field.

METHOD OF QUANTITATIVE TEST OF PROPORTIONS OF NUTRIENTS DIGESTED FROM FOOD BY MAN.

The method ordinarily used for testing the quantities of nutrients actually digested is, in substance, the following: The subject and the diet having been determined upon, samples of the food actually eaten

[1] Landw. Vers. Stat., 1889, 331; Ztschr. physiol. Chem., IX, 211; XI, 207, 529.

[2] Ztschr. Biol., 1879, 115, etc.

[3] Arch. Hyg., 1893, 626.

[4] König, Chem. d. mensch. Nahr. und Genuss., I, 1889, 46.

are analyzed, i. e., the nitrogen, fat, ash, etc., are determined by the usual methods of food analysis. If at the start it is not possible to provide enough of a certain food to last through the experiment, each fresh portion is analyzed.

It is desirable, though not absolutely necessary, that the length of the trial be not less than three days. This has been the length of most of the experiments in the Munich laboratory. The longer the period of the experiment the less will be the error due to imperfect separation of the undigested residue, and the more fully may the effect of the previous food be assumed to be eliminated. On the other hand, if the experiment is continued too long, the monotony of the diet may interfere with the abundant secretion of the digestive juices or otherwise tend to render the results abnormal. The accuracy of tests by this method depends in large part upon the accuracy of the separation of the feces which pertain to the food under investigation from that due to food which precedes and follows it. Two methods of separation are in general use. In one, the subject takes no food but milk during the twenty-four hours, or thereabouts, immediately preceding the actual test. The feces due to milk are whitish in color and of characteristic texture. It is not difficult to mechanically separate this portion from that due to the food which is taken during the test period. This is followed by another day of milk diet, and thus the digested residue of the food of the experimental period is separated from that of the succeeding food. The second plan consists in giving a teaspoonful, or less, of lampblack or powdered charcoal, preferably in gelatin capsules, with the last meal eaten before and the first meal after the period of the experiment. This imparts to the solid excrement a very dark color, and the mechanical separation is comparatively simple. A still more effective way to facilitate accurate separation is to take lampblack with milk before and after the test period. Full details of mechanical manipulations, as improved by later experiments, have been given by Prausnitz.[1]

The solid excrement is best collected in weighed dishes of porcelain or metal with smooth surface. It is dried, weighed, and determinations of water, nitrogen, ether extract, and ash are made by the usual methods.

The errors due to the imperfections of the methods of analysis and the presence of other materials than undigested residue will be better appreciated by considering what are the substances to be dealt with. The feces contain (1) the undigested residue of the food and (2) metabolic products.

The undigested residue consists of fragments of muscular fiber, tendon, ligament, elastic fiber, blood vessels, so-called "nucleins" (i. e., undigested proteid material), chlorophylloid matter, vegetable fiber, granules of starch, masses of fat, calcium and magnesium salts of fatty

---

[1] Ztschr. Biol., 1894, 335.

acids, magnesium ammonium phosphate, and cleavage products of proteids, including skatol and indol.

The metabolic products include epithelium mechanically separated from and mucus secreted by the walls of the alimentary canal, and residues of the digestive secretions. The chief of these latter are those coming from the salivary, peetic and pancreatic glands, and the bile. Pepsin and trypsin furnish considerable nitrogenous matter. The bile furnishes the bile acids and probably coloring matters. Whether cholesterin, which is normally present, comes from undigested residue or metabolic products or from both is not definitely known.

The quantity of the undigested residue is of course quite considerable, and varies with the conditions, including the quantity of food eaten, and very likely with the time the given diet is followed. For instance, it was found by Berthé[1] that when he consumed 60 grams of cod-liver oil per day the feces of the first day contained 8 grams ether extract, that of the seventh day contained 12 grams, and the quantity increased until on the thirteenth day it was 49 grams. It would seem that in this case the digestive organs gradually lost the power of assimilating large quantities of the materials soluble in ether. Rubner[2] found that in experiments of three days each, when 233 grams of butter and 145.8 grams of bacon, containing 95.6 per cent of fat, were eaten per day the feces contained 44.6 grams ether extract. When 240 grams of butter were eaten without the bacon only 15.2 grams ether extract was found in the feces, i. e., for large quantities of fat the quantity of ether extract in the solid excrement was proportionally larger than when small quantities were consumed.

The inorganic matter in the ash belongs to both the undigested residue and the metabolic products.

It is evident that where the methods of analysis of feces are the same as those commonly used for food the actual error must be even greater than the results imply. Only a portion of the nitrogen of either undigested residue or metabolic products is in the form of proteids. Much of the rest occurs in cleavage products of the proteids, the nature and amount of which is uncertain. Multiplying the nitrogen by 6.25 is far from giving the actual amount of either proteids or of total nitrogenous substance. The ether extract may contain all the neutral fats and probably more or less of cleavage products, materials from the bile and coloring matter. Ether will not remove these perfectly nor will it alone remove the fatty acids which are in combination with calcium and magnesium. The determinations of crude fiber are very likely as near the truth as in food analysis, but the quantities are small in both in experiments with men except when large amounts of coarse bread or similar food is eaten. Carbohydrates are estimated by difference. Such an estimate must include the combined errors of all the others except in so far as the latter compensate each other.

---

[1] Hermann's Handbuch der Physiologie, Vol. V, II, p. 243.

[2] Ztschr. Biol., 1879, 170-180.

The error due to the metabolic products outside of that from imperfect methods of analysis may be very considerable. It falls chiefly upon the protein and fat. The measure of digested nitrogen is the remainder after subtracting the nitrogen in the feces from the total nitrogen in the food. But the subtrahend includes both undigested residue and metabolic products. The latter products represent food material which has been previously digested and metabolized. All the metabolized nitrogen is, therefore, digested nitrogen, and it should be subtracted from the total nitrogen of the feces. The remainder would represent the nitrogen of the undigested residue. This last subtracted from the total in the food would leave the real quantity of nitrogen digested. The figures for the digestibility of nitrogen as ordinarily computed from experiments are really too small by the amount of the metabolic nitrogen in the feces.[1] In Table 12, beyond, the so-called "undigested nitrogen" in the meat, fish, and eggs made from 1.6 to 11.2 per cent of whole. It is customary to assume that this nitrogen belongs to the metabolic products and that the protein is completely digestible. The quantity of ether extract in the feces may be quite considerable because of the presence of bile acids and like products. When the quantity of fat in the foods is very large, as is apt to be the case with meats, the error in results due to the ether extract of the metabolic products may perhaps be neglected; but with vegetable foods in which the total fat is very small the error due to ether extract of metabolic products may seriously impair or entirely vitiate the results. The figures given for undigested fat of vegetable food in the tables beyond can not be taken as reliable.

It is often desirable to know whether the body is gaining or losing nitrogen with the diet used in a digestion experiment. For this purpose the urine is collected and the nitrogen in it is determined.

RESULTS OF EXPERIMENTS UPON QUANTITIES OF NUTRIENTS DIGESTED FROM DIFFERENT FOODS.

With the preceding statements the figures of Tables 11 and 12 will be clear.

Table 11 shows the general character and results of the experiments thus far reported upon the digestibility of food materials by man. It includes the final details of some 35 separate trials and the averages of some of these and other similar ones. The data are selected as fairly representative of the work of this kind reported up to the present time. Table 12 recapitulates experiments upon the digestibility of crude fiber (cellulose) of bread and vegetables by man.

[1] See investigations on this subject by Rieder, Ztschr. Biol., 1884, 378.

TABLE 11.—*Results of experiments on the digestibility of food materials by men; proportions of nutrients actually digested.*

No. of experiment.	Date of experiment.	Food used (quantities per day in grams).	Ingredients of food. Fresh substance.	Water-free substance.	Total nitrogen.	Ether extract.	Carbohydrates.	Ash.	Ingredients of feces. Fresh substance.	Water-free substance.	Total nitrogen.	Ether extract.	Carbohydrates.	Ash.	Food ingredients undigested. Water-free substance.	Total nitrogen.	Ether extract.	Carbohydrates.	Ash.
			Grams.	*Grams.*	*Grams.*	*Grams.*	*Grams.*	*Grams.*	*Grams.*	*Grams.*	*Grams.*	*Grams.*	*Grams.*	*Grams.*	*P. ct.*	*P. ct.*	*P. ct.*	*P. ct.*	*P. ct.*
1	1876	Beef, roasted	884	366.8	48.8	20.9	....	18.6	65.3	17.2	1.2	4.4	....	2.8	4.7	2.5	21.1	......	15
2	1876	....do	738	306.4	39.8	23.9	....	15.2	53.2	17.1	1.1	4	....	3.2	5.6	2.8	17.2	......	21.2
3	1884	Beef, roasted and boiled	798	193.7	21.3	48.5	....	11.8	20	5.4	.4	.9	....	1	2.8	1.6	1.8	......	8.2
4	1883	Beef, 1,200 grams; butter, 30 grams	1,230	291.2	38.5	61.4	....	13.3	62.3	13.8	.97	....	....	2.9	4.3	2.5	5.2	......	21.5
5	......	Average of Nos. 1, 2, 3, and 4	....	....	....	....	....	....	....	....	....	....	....	....	4.4	2.3	11.3	......	16.5
6	1883	Fish, haddock, 1,584 grams; butter, 30.5 grams	1,010.5	296.3	45.6	35.1	....	18.2	57.8	16	.9	3.2	....	4.1	4.9	.2	9	......	22.5
7	1876	Eggs, hard boiled	948.1	247.4	20.7	103.3	....	10.4	....	11	.6	5.2	....	1.9	5.2	2.9	.5	......	18.4
8	1876	Milk	2,438	315	15.4	95.1	102.4	17.8	96.3	24.8	1	4.7	....	8.7	7.8	6.5	3.3	......	48.8
9	1880	....do	1,790	224	10.6	53.7	91.3	....	69	15.0	.6	1.5	....	....	7.1	5.4	2.8	......	......
10	1889	....do	3,162.9	350.7	13.3	111.9	319.8	23.4	....	31.4	1.1	5.7	....	8.7	9	11.2	5.1	......	37.1
11	......	Milk. Average of 7 experiments, including Nos. 8, 9, and 10	....	....	....	....	....	....	....	....	....	....	....	....	8	7.6	4.5	......	45.1
12	1877	Milk, 2,291 grams; cheese, 200 grams	2,491	419.8	24.1	138.6	96.2	27.5	98.3	25.3	.9	3.8	....	7.2	6	3.7	2.7	......	26.1
13	......	Milk and cheese. Average of 3 experiments, including No. 12	....	....	....	....	....	....	....	....	....	....	....	....	8	3.8	7.3	......	37.5
14	1878	White wheat bread	689	421.4	7.6	......	391.1	9.9	95.2	23.5	2	....	....	2.5	5.2	25.7	......	1.4	25.4
15	......	White wheat bread. Average of 3 experiments, including No. 14	....	....	....	....	....	....	....	....	....	....	....	....	4.8	21.4	......	1	24.3
16	1869	Rye bread, 816.7 grams; butter, 50 grams	816.7	438.1	10.5	....	....	18.1	310.1	44.2	2.3	....	....	5.5	10.1	22.2	......	......	30.5
17	......	Rye bread. Average of 3 experiments including No. 16	....	....	....	....	....	....	....	....	....	....	....	....	12.2	28.9	......	......	34.9
18	1892	Wheat and rye bread with yeast	1,000	616.3	14.3	....	....	21.6	....	44.5	2.6	....	....	7.2	7.2	17.8	......	......	33.3
19	1892	Wheat and rye bread leavened	1,000	619.1	13.9	....	....	18.7	....	48.3	2.9	....	....	6.8	7.9	19.6	......	......	36.4
20	1877	Potatoes and butter	3,276.5	967.7	11.5	143.8	718.1	64	635	93.8	3.7	5.3	....	10.1	9.4	32.2	3.7	7.6	15.8

TABLE 11.—*Results of experiments on the digestibility of food materials by men; proportions of nutrients actually digested*—Continued.

No. of experiment.	Date of experiment.	Food used (quantities per day in grams).	Ingredients of food.						Ingredients of feces.						Food ingredients undigested.				
			Fresh substance.	Water-free substance.	Total nitrogen.	Ether extract.	Carbohydrates.	Ash.	Fresh substance.	Water-free substance.	Total nitrogen.	Ether extract.	Carbohydrates.	Ash.	Water-free substance.	Total nitrogen.	Ether extract.	Carbohydrates.	Ash.
			Grams.	*Grams.*	*Grams.*	*Grams.*	*Grams.*	*Grams.*	*Grams.*	*Grams.*	*Grams.*	*Grams.*	*Grams.*	*Grams.*	*P. ct.*	*P. ct.*	*P. ct.*	*P. ct.*	*P. ct.*
21	1878	Carrots and fat	....	411.6	6.5	47.3	281.9	41.2	1,092.6	85.1	2.5	3.1	....	14	20.7	39	6.4	18.2	33.8
22	1878	Wirsing (Savoy cabbage)	....	493.2	13.2	87.8	247.4	73.3	....	73.4	2.4	8.2	....	14.2	14.9	18.5	6.1	15.4	19.3
23	1876	Green beans, 540.2 grams; butter, 53.4 grams	593.6	101.1	1.4	46.2	25.5	11.9	57.4	15.2	.7	3.9	3.9	2.7	15	......	8.5	15.4	22.8
24	......	Green vegetables. Average of Nos. 22 and 23	....	....	....	....	....	....	....	....	....	....	....	....	15	......	7.3	15.4	21
25	1879	Peas, dry	600	521.1	20.4	7	357	30.1	260.1	48.5	3.6	5	....	8.2	9.1	17.5	63.9	3.6	32.5
26	......	Average of 3 experiments, including 25	....	....	....	....	....	....	....	....	....	....	....	....	11.2	19.7	83.5	4.9	36.5
27	1876	Lentils, dry	....	223.5	8.7	....	....	....	....	....	3.5	....	....	....	......	40.2	......	......	......
28	......	Beans, 500 grams; fat, 25 grams; flour, 17 grams; salt, 18 grams	560	494.2	17.9	....	....	17.9	106.4	....	5.4	....	....	9.8	18.3	30.3	......	......	28.3
29	1884	Maize meal	629	540.9	6.9	10.1	481	18.4	....	34.1	1.3	4.2	10.4	5.6	6.3	18.3	42.1	3.4	30.5
30	1884	Maize meal and butter	871.3	749.7	8.6	92.5	590.3	21.3	....	53.2	2.7	7.1	21.8	7.7	8	31.5	56.8	3.7	35.9
31	1884	Maize meal and cheese	924.3	774	14.4	37.3	612.5	37.8	....	32.5	1.1	4.4	14.1	7.3	4.2	7.3	9.3	2.3	19.4
32	1876	Maize meal, butter, meat extract and cheese	865.3	738	14.7	48.6	....	26.8	198	49.3	2.3	8.5	....	8	6.7	15.5	17.5	3.2	30
33	......	Average N. of 28 and 29. Carb. of 28, 29, 30 and 31	....	....	....	....	....	....	....	....	....	....	....	....	......	24.9	......	3.1	......
34	1877	Rice, 638 grams; fat, 82.3 grams; meat extract, 24.4 grams	759.8	660.2	10.4	74.1	492.9	23.8	194.6	27.2	2.1	5.2	....	3.6	4.1	20.4	7.1	.9	15
35	1883	Bread, white, 400 grams; peas, 80 grams; potatoes, 200 grams; condensed milk, 70 grams; egg albumen, — grams; sugar, 20 grams; cheese, 30 grams; beer, 330 grams; butter, 72 grams=60.5 fat	....	....	16.5	73.3	402	....	107	....	....	1	....	....	......	......	1.3	......	......
36	1883	Bread, milk, etc., as above. Artificial butter, 70 grams=63 fat	....	....	16.5	75.8	402	....	148	....	....	2.5	....	....	......	......	3.3	......	......
37	......	Average of 3, natural butter, including No. 35	....	....	....	....	....	....	....	....	....	....	....	....	......	......	1.6	......	......

38	......	Average of 3, artificial butter, including No. 36	....	....	....	....	....	....	....	....	....	....	....	....	....	....	3.6	....	....
39	1877	Beef, 614 grams; bread, 450 grams; bacon, 100 grams	1,164	....	23.6	99	259.6	23.5	209.1	46.5	2.9	17.2	....	6.7	8.5	12.1	17.4	1.6	28.5
40	1886	Potatoes, 1,700 grams; suet, — grams; beer, — grams; gluten, 200 grams; (= 24.5 N.)	....	....	31.7	101	379.6	18.3	159.3	25.9	2	2.6	1.4	3.1	....	6.4	2.6	.4	16.9
41	1888	Rice, boiled, 1,200 grams; salted radish, 100 grams; vegetables, etc., 350 grams; fish, 150 grams	1,800	615.7	17.5	19.5	461.2	21.2	97.9	22.2	2.2	4	4	3.5	3.6	12.6	.8	.8	16.5
42	1893	Soup, macaroni, vegetables, white bread, potatoes, meat, etc., (average of 8 experiments)	....	471.1	11.2	31.9	368.9	....	....	32	2.2	3.9	15	....	6.8	18.6	12.4	4.1	....

NOTE.—The years are, as nearly as we can find in the literature at our disposal, those in which the experiments were made. Ash, average of 5 experiments.

1, 2. Rubner, Ztschr. Biol., 1879, 121, 125.
3. Malfatti, König's Chem. d. mensch. Nahr. und Genuss., 1889, Vol. I, p. 37.
4, 6. Atwater, Ztschr. Biol., 1888, 16.
7, 8. Rubner, Ztschr. Biol., 1879, 128, 130.
9. Camerer, Ztschr. Biol., 1880, 493.
10. Prausnitz, Ztschr. Biol., 1889, 533.
12, 14. Rubner, Ztschr. Biol., 1879, 136, 151.
16. Meyer, Ztschr. Biol., 1871, 916.
18. Prausnitz, Ztschr. Biol., 1894, 328.
20, 21, 22. Rubner, Ztschr. Biol., 1879, 147, 166, 163.
23, 25. Rubner, Ztschr. Biol., 1880, 127, 128.
27. Strumpell, König, loc. cit., 46.
28. Prausnitz, Ztschr. Biol., 1890, vol. 26, 227.
29, 30, 31. Malfatti, König, 1889, 46.
32, 34. Rubner, Ztschr. Biol., 1879, 141, 144.
35, 36. Mayer, Landw. Vers. Stat., 1883, vol. 29, 215.
39. Rubner, Ztschr. Biol., 1879, 170.
40. Constantinidi, Ztschr. Biol., 1887, 447.
41. Mori, Ztschr. Biol., 1889, 117.
42. Manfredi, Arch. Hyg., 1893, 576.

TABLE 12.—*Crude fiber (cellulose) digested in experiments with men.*

No. of experiment.	Date of experiment.	Food used, (quantities per day in grams).	Water-free substance in food.	Crude fiber— In food.	In feces.	Undigested.
			Grams.	*Grams.*	*Grams.*	*Pr. cent.*
40	1886	Potatoes, 1,700 grams: fat, 100 grams; beer, 500 cubic centimeters; gluten, 200 grams	....	5.6	1.2	22
43	1886	Potatoes, 1,700 grams; fat, 100 grams; beer, 500 cubic centimeters	....	4.8	1	21.1
44	1869	Celery, cabbage, carrots, total 1,050 grams	139	12.5	4.7	37.3
45	1869	Celery, cabbage, carrots, total 883.3 grams	117.8	10.4	5.5	52.7
46	1886	Bread, fruit, dry fruit, oil, total 1,802 grams	719	16	9.2	56
47	1886	Bread, fruit, dry fruit, oil, total 1,764.8 grams	692.5	16.7	6.3	37
48	1888	Rice and barley, salted radish, vegetables, total 2,150 grams	523.8	17.4	4.2	24
41	1888	Rice, salted radish, vegetables, etc., fish, total 1,800 grams	615.7	4.6	.8	17.5
49	1888	Rice, salted radish, vegetables, etc., meat, total 1,500 grams	580.1	6	.5	8.6
18	1892	Wheat and rye bread (with yeast) 1,000 grams	616.3	4.3	2.5	63.1
50	1892	Wheat and rye bread (with yeast), 900 grams	554.6	3.9	2	50.1
19	1892	Wheat and rye bread (leavened), 1,000 grams	619.1	4.7	3.3	70
51	1892	Wheat and rye bread (leavened), 900 grams	557.2	4.3	1.5	36.4
52	1892	Decorticated (entire) rye bread (with yeast), 1,000 grams	647.3	8.2	3.7	45.2
53	1892	Decorticated (entire) rye bread (with yeast), 900 grams	582.6	7.4	4.1	55.9
54	1892	Decorticated (entire) wheat bread (with yeast), 1,000 grams	640.8	6.7	3.7	55.4
55	1892	Rye bread (with yeast), 1,000 grams	623	8.2	4.9	59.7
56	1892	Rye bread (with yeast), 900 grams	570	7.6	4.8	63.9
57	1892	Wheat bread (with yeast), 1,000 grams	610.2	8.9	4.2	47.4
58	1892	Wheat bread (with yeast) 900 grams	576.1	8.1	3.8	46.6

44, 45. Weiske, Ztschr. Biol., 1870, 456.
46, 47. Voit, Ztschr. Biol., 1889, 261, 274.
48, 49. As 41 above. In 49, 200 cubic centimeters of milk were used per day.
50 to 58 (*see* 18, 19).

NOTES ON TABLES 11 AND 12.

Experiments with beef. Four experiments are here quoted. Nos. 1 and 2 by Rubner, No. 3 by Malfatti, and No. 4 by Atwater. Nos. 1 and 2 were carried out in July and June, 1876, in the laboratory of the Physiological Institute of the University of Munich. The subject was a medical student 22 years old, weighing 72 kilograms. The time of each was 3 days.

In No. 1 the food was lean beef, which was prepared by separating the fat, gristle, and connective tissue as completely as was practicable with shears. The meat thus prepared was fried or roasted with a little butter, onion, salt, and pepper, and eaten either with well water or carbonated water as a beverage. For purposes of analysis specimens of the meat after it had been cooked with the above materials were taken each mealtime and fat and water determined. The quantity of nitrogen in the meat was estimated. For this estimate the nitrogen content of the dry, fat-free flesh was assumed to be 14.11 per cent.[1] The ash was estimated. In the feces the nitrogen was determined by the soda-lime method. Dry matter, fat, and ash were determined by the ordinary methods. The separation of the feces was effected by a milk diet on the day before and the day after the period. No. 2 was made with the same person and by the same methods. Although the meat was extremely palatable, it was almost impossible for the subject to eat it on the third day. Eating meat alone caused a strong aversion to it.

No. 3 is as quoted by König from the original publication.[2]

Nos. 4 and 6 were made by Atwater for a comparison of the digestibility of meat and fish. The work was done in the Munich Physiological Laboratory in 1883.

[1] As the result of early analyses Voit and his pupils in the Munich Physiological Laboratory assumed for a number of years that lean beef, from which the visible fat and the larger portions of tendon and other connective tissue had been removed as completely as could be done with shears, had an approximately uniform composition.

[2] "Sitzungsber. Wien. Akad. Wissensch., 1884, III Abth. Dez.-Heft."

The subject was a medical student, some 22 years of age and weighing 70 kilograms. The nitrogen, fat, dry matter, and ash were determined in each case by the usual methods. The meat was purchased in a market in Munich, and the separations of fat and connective tissue were made by the same laboratory servant and in the same way as in other experiments in this laboratory. The fish was haddock brought from the Baltic and sold in Munich. The experiments were made in winter, and the fish was well preserved. The fish was freed from skin, bone, etc., by the method followed by Atwater and assistants in the chemical investigation of fishes.[1] The composition of the flesh agreed very nearly with that of haddock taken off the New England coast. The beef was leaner in appearance and contained considerable less fat than has been found in any one of a number of specimens purchased in Middletown, Conn., and treated in the same way. There is, however, no reason to assume that this difference materially affects the digestion.

No. 7. This experiment was made in 1876 by Rubner, also in Munich. The subject was a student of medicine, 24 years of age, weighing 46 kilograms. The time was 2 days. The only food was eggs, which were boiled hard in the shells and eaten with a little salt. Water was the only beverage used. The dry matter in the eggs was determined, but the nitrogen, fat, and ash were calculated from previous analyses by Voit. Determinations were made of the dry matter, nitrogen, fat, and ash in the feces. Milk was not used to separate the feces, as they were characteristic. It will be seen that practically the same amount of dry substance and nitrogen were digested in the experiment with eggs as in those with meat, while the percentage of digested fat in the eggs was larger. Rubner remarks:

"It is well known that the protein of eggs is coagulated by heating to 100°, while the syntonin and myosin, the chief protein in meat, is not coagulated by heat. It requires however, an acid, or the action of digestive juices, to dissolve it. From the fact that eggs are as completely digested as meat it does not follow that they are digested in the same time or that the hard-boiled eggs do not produce more disturbance in the digestive organs. It is highly probable that soft-boiled eggs are as thoroughly digested as hard boiled."

No. 8. This experiment was also made by Rubner in Munich in 1876. The subject was a professional man 27 years of age, weighing 71 kilograms. The time was 3 days. The only food used was milk. The dry matter, nitrogen, fat, sugar, and ash in the milk were estimated from previous analyses by Voit. The feces were analyzed, the dry matter, nitrogen, ether extract, and ash being determined. "No carbohydrates or protein were found." The milk feces could, of course, be easily separated.

Rubner remarks that the solid matter of milk is not as completely digested by adults as that of meat or eggs. This is largely due to the fact that the percentage of ash in the solid matter is larger in milk than in the other two. According to Rubner the undigested organic matter from meat amounted to 4.1 to 4.7 per cent, from eggs 4.7 per cent, and from milk 5.4 per cent of the whole—not a very considerable difference. Young children digest milk more completely than adults. Forster[2] found that only 6.4 per cent of the solid matter of milk was undigested by a nursing infant. This may, perhaps, be explained by the fact that a considerable part of the ash of milk is composed of calcium salts, and these would be more needed by the young organism for the formation of bones than would be the case with an adult, and not so much would be left in the undigested residue to form insoluble salts of the fatty acids.

No. 9 was made by Camerer, in 1880, in Reidlingen. The subject was his daughter, 12 years of age, and weighing 26.3 kilograms. The time was 4 days. It was the investigator's intention to make the time 6 days, but the milk became so distaste-

[1] The Chemical Composition and Nutritive Value of Food Fishes and Invertebrates, by W. O. Atwater. Report U. S. Commissioner of Fish and Fisheries, 1888, pp. 679–868. Washington, 1891.

[2] Ztschr. Biol., 1879, 135.

ful that it was discontinued. A little coffee was used with the milk, but is not taken into account, as the quantity of solid matter would have been very small. In the original account nothing is said of the separation of the feces. Analyses of dry matter, casein, fat, and ash of the milk, and of the dry matter, nitrogen, fat, and ash of feces, were made in the laboratory of Hüfner, in Tübingen.

No. 10 was made by Prausnitz in the laboratory of the Physiological Institute in Munich, in 1889. The subject was the laboratory servant who had been so often used in Rubner's experiments. His weight was 74 kilograms. The time was 3 days. The only food was milk. This was purchased in quantity, thoroughly mixed, put into little flasks, and sterilized by heating for 2 hours in a Koch steam sterilizer. The milk was kept in the flasks on ice, and slightly warmed before it was used. The separation of the feces was made by eating meat before and after the experiment. Determinations of dry matter, casein, total protein, nitrogen, fat, sugar, and ash in the milk were made; and of dry matter, nitrogen, ether extract, and ash in the feces.

Prausnitz's work bears out Rubner's statement that the amount of matter undigested by adults is larger in milk than in any other animal food. As Prausnitz says, "The presence of larger quantities of nitrogen in the milk feces does not of necessity mean that the protein of milk is not all digested. The nitrogen might be derived from digestive juices excreted with the undigested residue. This consideration does not affect the value of milk as a food. It is one of the most convenient, useful, and at the same time inexpensive sources of protein." He urges its use by the poorer classes.

No. 12 was made by Rubner in Munich, in 1877. The subject was a soldier, 23 years old, weighing 72 kilograms. The time was only 1 day. Rubner wished to make the experiment with cheese alone, but could find no one willing to live upon cheese without other food; therefore milk and cheese were used together. The cheese used was Allgäuer (similar to what is called "Swiss" cheese in the United States), and the dry matter, nitrogen, fat, and ash in it were estimated from analyses of cheese by Forster. The feces were analyzed. The separation was of course easily made. It is very noticeable that, though the quantity of solid matter, nitrogen, fat, and ash in the food was much larger than in the experiment of Rubner with milk alone, the quantity of solid matter, nitrogen, ether extract, and ash in the feces was not large. Rubner considers that this may be due to the fact that when milk is the sole article of diet its casein forms large masses in the stomach. The mechanical effect of the small pieces of cheese is to make these lumps of casein smaller, and hence more easily acted upon by the digestive juices.

No. 14 was made by Rubner in Munich, in 1878. The subject was the laboratory servant above mentioned. He was then 43 years of age, and weighed 74 kilograms. The time was 2 days. The food was bread made from fine wheat flour. The dry matter, nitrogen, and ash in the flour and yeast were determined by analysis, and the results were used in computing the composition of the bread, which was made in the physiological institute where the experiments were carried on. The feces were analyzed. The method of separation of the feces was not stated, but presumably milk was used before and after the experiment. The quantity of the solid excrement was not large, but contained more nitrogen than that from animal foods. The nitrogen balance showed that the subject lost nitrogen at the rate of 3.4 grams per day.

No. 16 was made by Meyer in the same laboratory, in 1869. The subject was a healthy young man. The age and weight were not stated. The time was 4 days. The food was rye bread, made in Munich. Some coarse wheat flour was mixed with the rye, and the bread was leavened. To make the bread more palatable, 50 grams of butter were eaten with it, and 2 liters of beer per day were used as a beverage. The separation of the feces was made by a diet of meat before and after the experiment. Dry matter, nitrogen, and ash were determined in the bread and feces. The subject had a feeling of hunger during the experiment, though the quantity of bread eaten was rather large.

Nos. 18 and 19 were made by Menicianti and Prausnitz in the same laboratory, in 1892. The time of each was 3 days. The subject was a physician, 25 years old, weighing 85 kilograms. In No. 18 the food was entirely bread made of wheat and rye flour with yeast. In No. 19 the bread was made of part of the same flour, but was leavened. Full analyses of the bread and feces were made. The separation of the feces was made with milk.

The flour was ground especially for the investigators, and the bread made under their supervision. The utmost care was taken in all details.

The object of the two investigations last named was to observe the effect of the different processes of bread making upon digestibility. The process of leavening is still much used in Europe. Dough is kept from one baking until the next, and this "leaven" is used instead of yeast. An opinion prevails in England that leavened bread is not as palatable as bread made by other processes.[1] The leavened bread in this experiment gave a larger undigested residue than that made with yeast. In Prausnitz's opinion this may be due to the fact that the presence of larger quantities of bacterial cleavage products in the leavened bread caused a greater excretion of intestinal juices and this effected the increase in weight of the metabolic products in the feces.

No. 20 was made by Rubner in Munich in 1877. The subject was a Bavarian soldier who was accustomed to a diet consisting largely of potatoes. The time was 2 days. The food consisted of potatoes, which were boiled and eaten with salt or butter, or with oil and vinegar as salad. It is not stated whether analyses were made of the potatoes or whether the composition was estimated from other analyses. Analyses of dry matter, nitrogen, ether extract, and ash in the feces were made. The separation of the feces was by means of a milk and cheese diet. The nitrogen was poorly digested and the body lost nitrogen at the rate of 1 gram per day.

Nos. 21 and 22 were made by Breuer in the Munich laboratory in 1878, and reported by Rubner. Breuer was himself the subject; his age and weight are not stated. The time in No. 21 was 2 days; in No. 22 it was 3 days. The vegetables, carrots in No. 21, Savoy cabbage in No. 22, were in each case the only food. The same method was followed in the preparation of both; the fresh vegetable was cooked in water with salt and fat for one-half hour in a covered vessel. The separation of the feces was by means of a milk diet. Analyses of the vegetables and of the feces were made. It was the intention of the investigator to continue the carrot diet for 3 days, but it became so distasteful that this could not be done. In the case of the carrots the fat was well digested and the carbohydrates very poorly. In each case the body lost nitrogen; with carrots 8.5 grams per day, and with Savoy cabbage 6.9 grams per day.

No. 23 was made by Rubner in Munich in 1876. The subject was a medical student 22 years old. The time was 2 days. The food was fresh green beans, (presumably "string beans") which were cooked in water with some butter. Whether analyses of the beans were made or not is not stated, but it seems probable that they were. The feces were analyzed. Rubner remarks that too much value should not be placed on this experiment, since the quantity of solid matter in the diet was too small to serve in any adequate manner as food.

No. 25 was made by Rubner in Munich in 1879. The subject was the laboratory servant above mentioned. The time was 2 days. The food consisted of peas which were purchased dry, carefully cleaned, and cooked in water 2 or 3 hours. They were then soft and were pressed through a fine sieve. Salt was eaten with the peas and 1 liter of beer per day was used as a beverage. Full analyses of the peas and of the feces were made. The chief point of interest in this investigation is the digestibility of the nitrogen. In common with other vegetable foods peas contain a large percentage of undigested nitrogen, but they rank among the best of vegetables as a furnisher of nitrogen.

---

[1] Jago. Chemistry of Wheat Flour and Bread and Technology of Breadmaking, p. 328.

No. 27 was made by Strümpell, who was himself the subject, and is quoted from König.[1] The food consisted of lentils, a very common food material in Europe. They were soaked in water and then boiled. The nitrogen was much less thoroughly digested than in dry peas or beans, but it seems probable that if more experiments were made the nitrogen would be found to be more digestible.

No. 28 was made by Prausnitz in the Munich laboratory (in 1889?). The subject was the laboratory servant mentioned above. The time was 3 days; the food white beans. These were soaked in water overnight and then cooked in salted water until soft. Some flour was browned in fat and this mixed with the beans, with addition of a little vinegar and some of the water in which the beans were cooked. Analyses were made of the beans and of the feces. The chief interest in this experiment attaches to the digestion of the nitrogen. The amount undigested, 30.3 per cent, is much larger than in the experiment with peas; but it must be remembered that the peas were eaten in the form of a mush, while the beans were, for the most part, whole. This might have a considerable influence on the digestibility.

Nos. 29, 30, and 31 were made by Malfatti[2] and are quoted from König. The subject was a man 20 years old weighing 75 kilograms. The food in No. 29 was maize meal cooked into a mush with water. In No. 30 butter and in No. 31 cheese was added to the meal. Maize meal cooked in this way either with or without butter or cheese is much used in northern Italy under the name of polenta. It forms almost the sole food of a large class of the poor people. The quantity of nitrogen in the maize meal is not large but it is very well digested.

No. 32 was made by Rubner in Munich in 1876. The subject was a medical student 22 years old, weighing 72 kilograms. The time was 2 days. The food was maize meal cooked to a mush with water and butter, with addition of grated Parmesan cheese. The subject of the experiment did not relish the food thus prepared and after the first meal made of it, meat extract was added; 1,250 centimeters of beer daily were used as a beverage. The composition of the food was estimated from König's compilation of analyses. The feces were analyzed. The digestibility of the nitrogen, fat, and ash agree quite closely with No. 29. · The body lost 2.2 grams of nitrogen per day.

No. 34 was made by Rubner in Munich in 1877. The subject was the same as in No. 32. The time was 3 days. The food was rice which was cooked in water, or water and meat extract; a little fat and salt were added. Analyses of the food were not made, but the composition was estimated from the figures of König and others. The feces were analyzed. The separation of the feces was by means of a meat diet at the beginning and milk at the end of the experiment. The body lost 3.2 grams of nitrogen per day.

Nos. 35 and 36 were made by Mayer at Wageningen, in Holland, in 1882. The subject was a man 39 years old, weighing 70 kilograms. The time in each case was 1 day, but three of these trials of a day each were made with each material. The object of the experiments was to test the comparative digestibility of butter and oleomargarine. Each was eaten in a diet which was otherwise practically the same in all three trials. It consisted of bread, peas, potatoes, condensed milk, egg albumen, sugar, cheese, and beer, and contained little fat aside from the butter or oleomargarine. In No. 35, 72 grams of butter, containing 60.5 grams fat, and in No. 36, 70 grams oleomargarine, containing 63 grams fat, were eaten. The fat in the butter was determined. The nitrogen, fat, ash, and carbohydrates of the food were calculated from analyses. Ether extract of the feces was made. The butter came from a farm in Holland and the oleomargarine from a manufactory in The Hague. The natural butter was slightly better digested than the artificial. The same difference is seen in Nos. 37 and 38, which are averages each of three practically identical experiments. The little fat in the diet, aside from the two sorts of butter, comes

[1] König quotes from Centbl. med. Wiss., 1876, p. 47.

[2] Sitzungsber. Wien. Akad. Wissensch., 1884, vol. 110, III part.

largely from the condensed milk, and so is really the same as butter fat, and should not affect the results materially. It is very probable that the other extract in the case of butter was entirely composed of metabolic products, coloring matter, etc. The coefficient of digestibility of butter fat would then be 100 per cent. If the same amount of coloring matter were excreted in the experiment with oleomargarine its digestibility would be 98 per cent. For all practical purposes oleomargarine would thus be as digestible as butter.

No. 39 was made by Rubner, in Munich, in 1877. The subject was the laboratory servant mentioned above. The time was 2 days. The object of the experiment was to observe the quantity of fat which could be digested. The fat was given in a practically constant mixed diet consisting of bread and meat. Whether analyses of the food were made or whether its composition was calculated from known figures is not stated. The feces were analyzed. The separation was made by means of a milk diet. In No. 39, 100 grams of bacon (with 95.6 per cent fat) were eaten. The excreted fat was 17.2 grams, or 17.4 per cent of the amount consumed. Three other experiments, with larger quantities of fat, were made. These are tabulated with No. 39 for convenience.

Comparative digestion of fat when taken in different quantities.

No.	Kind of fat eaten.	Fat in food.	Ether extract in feces.	Ether extract in feces.
		Grams.	*Grams.*	*Per cent.*
39	Bacon (95.6 per cent fat)	100	17.2	17.4
a	Bacon (95.6 per cent fat)	200	15.2	7.8
b	Butter	240	5.8	2.7
c	Bacon (95.6 per cent fat)	145.8	44.6	12.7
	Butter	233		

It will be seen that when a large quantity of fat was consumed the excreted quantity was large. It is also noticeable that butter was more digestible than bacon. Rubner remarks that this is very likely due, in part, to the fact that the butter is more thoroughly acted upon by the digestive juices, since it is in a finely divided condition. The feces contained small pieces of bacon which had evidently not been disintegrated and were unchanged. The fat in the bacon is also inclosed in fat cells, and hence less easily attacked.

No. 40 was made by Constantinidi in the Munich laboratory in 1886. The subject was the laboratory servant mentioned above. The time was 3 days. The diet consisted of potatoes, cooked in water to which fat was added, and "gluten," a vegetable proteid compound made from waste products of wheat. Beer was used as a beverage. The food and excreta were analyzed. The separation of the feces was effected by means of milk. The object of the investigation was to see if the vegetable protein would furnish a fair substitute for animal protein, as ordinarily consumed in meat, or any other expensive protein substance. The nitrogen in the feces was only 6.4 per cent of the whole. The body made a daily gain of 3.6 grams nitrogen. When a second experiment was made like the above, but without the gluten, the body lost 2.3 grams of nitrogen per day. The gluten furnished, therefore, a valuable and cheap nitrogenous food.

No. 41 was made by Mori, in the laboratory of the Physiological Institute in Tokio, Japan. He was himself the subject; his age was 23 years, and his weight 52 kilograms. The time was 6 days. The diet consisted of boiled rice, bean cheese, vegetables (potatoes, etc.), salted radish, and fish. This is a typical diet of Japanese of the middle classes to-day near the coast, and before European influence was so pronounced in Japan, was very widespread among even the better classes. The food and excreta were analyzed. It will be seen that the food was well digested. The nitrogen balance showed a daily gain of 0.9 grams nitrogen by the body.

No. 42 was made by Manfredi, in the laboratory of the Hygienic Institute of the University of Naples, in 1892. The time was from 3 to 7 days in the different trials. The subjects were poor people in Naples. Eight studies were made. Full details of the food are given in the chapter on dietaries, page 195. The food and excreta were analyzed. The separation was made by means of grapes. The skins and seeds are excreted undigested and afford a means of identifying the food with which they were eaten, although the demarcation is less sharp than with lampblack or milk. Determinations of crude fiber in food and feces were also made.

Nos. 46 and 47 were made by Voit, in the Munich laboratory, in 1886. The subject of No. 46 was an upholsterer, 28 years old, weighing 57 kilograms. The time was 14 days, and divided into periods of 5, 5, and 4 days. The subject of No. 47 was a laboratory servant, weighing 74 kilograms. The time here was only 3 days, and the experiment can not be taken as of much value, as the food was so distasteful that it could only be eaten with difficulty. Details of No. 46 will be found under dietary No. 44, on page 192.

Nos. 48 and 49 were made by Mori, in Tokio, in 1888. Details of the diet will be found on pages 197 and 198.

No. 43 is included in the description of No. 40.

Nos. 50 to 58 were made by Prausnitz, in Munich, in 1892. They form a series with Nos. 18 and 19. The subject in Nos. 18, 19, 52, 54, 55, and 57 was a physician 25 years old, weighing 82 kilograms. The subject in Nos. 50, 51, 53, 56, and 58 was a laboring man 34 years old, weighing 85 kilograms. The time was in each case 3 days. As was stated under Nos. 18 and 19, the object of these experiments was to test the digestibility of various kinds of bread made with yeast and leavened. Full analyses of food and feces were made, including crude fiber. The quantity of crude fiber in the bread was small, 3.9 to 8.9 grams per day, and it was in general about half digested.

The attempt is made in Table 13, herewith, to epitomize the results of the experimental inquiry above detailed. The coefficients of digestibility of the nutrients, especially in the cases of protein and fat, are somewhat higher than the corresponding figures in Table 11, the reason being that allowance is made in Table 13 for the metabolic products. Of course such figures as these are only estimates. Probably more reliance can be placed upon the coefficients of animal than upon those of vegetable foods. The figures for the digestibility of the protein of vegetable foods are doubtless nearly accurate, those of the carbohydrates reasonably so, and those of the fats least reliable and in some cases of very little value.

TABLE 13.—*Digestibility of nutrients of food materials.*

Animal foods.	Percentages digested.			Vegetable foods.	Percentages digested.		
	Protein.	Fat.	Carbohydrates.		Protein.	Fat.	Carbohydrates.
	P. ct.	*P. ct.*	*P. ct.*		*P. ct.*		
Beef and veal	100	95	....	Wheat flour, fine	85	90 per cent assumed for all.	95 per cent assumed for all.
Mutton	100	95	....	Wheat flour, medium	81		
Pork	100	95	....	Wheat flour, coarse	75		
Fish and oysters	100	95	....	Rice	85		
Milk	100	96	100	Macaroni	85		
Cheese	100	95	100	Rye flour	78		
Butter	....	96	....	Maize meal	85		
Oleomargarine	....	95	....	Potatoes	75		
Tallow	....	95	....	Cabbages, turnips, etc	80		
Lard	....	95	....	Beans	85		
Oils	....	95	....	Peas	85		
Eggs	100	98	....				

Table 14 exhibits, for a number of food materials, the proportions of nutrients which are estimated to be digestible. It was computed by applying the figures in Table 13 to the figures for composition of food materials in Table 2, above. Thus, in beef the whole of the protein (100 per cent in Table 13) is assumed to be digestible. The average of the analyses of shoulder of beef, epitomized in Table 2, gives 19.5 per cent of protein. This is, accordingly, the percentage assumed to be digestible. In wheat flour 85 per cent of the total protein is estimated to be digestible, the remainder, 15 per cent, being undigestible. The wheat flour of Table 2 contains 11 per cent of protein. Accordingly 9.4 per cent will be digestible and 1.6 per cent undigestible.

TABLE 14.—*Estimates of proportions of digestible and undigestible nutrients in food materials.*

Food materials (edible portion).	Protein.			Fats.			Carbohydrates.		
	Digested.	Undigested.	Total.	Digested.	Undigested.	Total.	Digested.	Undigested.	Total.
	P. ct.	*P. ct.*	*P. ct.*	*P. ct.*	*P. ct.*	*P. ct.*	*P. ct.*	*P. ct.*	*P. ct.*
Beef:									
Shoulder	19.5		19.5	14.8	0.8	15.6			
Sirloin	18.5		18.5	19.5	1	20.5			
Round	20.5		20.5	9.6	.5	10.1			
Veal, shoulder	20.2		20.2	9.3	.5	9.8			
Mutton:									
Shoulder	18.1		18.1	21.3	1.1	22.4			
Loin	15		15	33.3	1.7	35.0			
Pork:									
Shoulder	16		16	31.2	1.6	32.8			
Very fat	.9		.9	78.7	4.1	82.8			
Eggs	14.9		14.9	10.3	.2	10.5			
Milk	3.6		3.6	3.8	.2	4	4.7		4.7
Butter	(?)	(?)	1	81.6	3.4	85	.5		.5
Cheese:									
Full cream	28.3		28.3	33.7	1.8	35.5	1.8		1.8
Skim milk	38.4		38.4	6.5	.3	6.8	8.9		8.9
Haddock	16.8		16.8	.3		.3			
Bluefish	19		19	1.1	.1	1.2			
Cod	15.8		15.8	.4		.4			
Shad	18.6		18.6	9	.5	9.5			
Mackerel	18.8		18.8	7.8	.4	8.2			
Halibut	18.3		18.3	4.9	.3	5.2			
Salmon	21.6		21.6	12.7	.7	13.4			
Wheat flour:									
Fine	9.4	1.6	11	.9	.2	1.1	71.1	3.8	74.9
Medium	9.5	2.2	11.7	1.4	.3	1.7	68.1	3.6	71.7
Cracked wheat	8.9	3	11.9	1.4	.3	1.7	70.9	3.7	74.6
Maize meal	7.8	1.4	9.2	3	.8	3.8	67.1	3.5	70.6
Rice	6.3	1.1	7.4	.3	.1	.4	75.4	4	79.4
Potatoes	1.6	.5	2.1	(?)	(?)	.1	17	.9	17.9
Turnips	1	.2	1.2	(?)	(?)	.2	7.8	.4	8.2
Beets	1.2	.3	1.5	(?)	(?)	.1	8.5	.4	8.9
Wheat bread	7.1	1.7	8.8	1.4	.3	1.7	53.4	2.8	56.2
Rye bread	6.5	1.9	8.4	.4	.1	.5	56.7	3	59.7
Graham bread	7.1	2.4	9.5	1.1	.3	1.4	50.6	2.7	53.3

FURTHER CONSIDERATIONS REGARDING DIGESTIBILITY.

The estimates of coefficients of digestibility above made, as well as the statistical data upon which they are based, apply to the quantities of nutrients digested by healthy individuals from wholesome food properly cooked and masticated. They do not in any way explain the reasons for the differences of digestibility of the nutrients in different kinds of food materials. They throw but little light upon the ease and time of digestion and the fitness of the digested matters for the user.

As to the reasons for the differences of digestibility, but comparatively little is definitely known. There are, however, certain a priori considerations which help toward explaining them. For example, the digestibility of the proteids is discussed by Prof. R. H. Chittenden, as follows:[1]

If of two foods possessing a like composition one be more easily digestible, that one, though containing no more available nutriment than the other, is in virtue of its easier digestibility more valuable as a food stuff, and in one sense more nutritious, as well as more economical for the system. The ease with which a proteid food can be utilized to supply the needs of the body depends, therefore, not only upon its intrinsic qualities, viz, whether it has been derived from the animal or vegetable kingdom, and upon its chemical and other peculiarities, but also, to a great extent, upon the way in which it has been prepared for digestion.

Proteid foods are rendered available for the needs of the body mainly through the action of the digestive juices. These convert the insoluble food stuffs into soluble and diffusible products, capable of being directly absorbed by the circulating blood. The two juices having this functional power are the gastric and pancreatic fluids, the former performing its work in the stomach and the latter in the small intestine. Both accomplish essentially the same object in a somewhat different manner, viz, the conversion of the insoluble proteids into soluble albumoses or proteoses and peptone, which are taken up from the alimentary tract by the blood and distributed to all parts of the body, where they may be utilized according to their several functions in nutrition. It is thus easily conceivable that while one class of proteid food stuffs may be somewhat resistant to the digestive action of the acid gastric juice, it may be more readily attacked by the alkaline pancreatic juice, or the reverse may be true. In any event, the full utilization of the ingested food stuff by the system depends in great part upon the completeness of its digestion by these two digestive fluids.

As a general rule the proteids of animal origin, as of beef, mutton, eggs, etc., are more easily digestible than those derived from the vegetable kingdom. This is partially due to the nature of the animal proteids; but another factor of even greater moment is the large admixture of extraneous matters ordinarily associated with vegetable proteids. Thus, oatmeal, wheat flour, potatoes, etc., contain a comparatively small amount of proteid matter, admixed with a large bulk of starchy material and some cellulose. Hence, in a purely vegetarian diet, a large bulk must be consumed in order to obtain even the minimum amount of proteid food required. In other words, the nitrogenous or proteid matter is so diluted by large masses of cellulose and starch that an excess of work is thrown upon the alimentary organs, which not only causes discomfort, but is a physiological loss, entailing the working over by the system of large quantities of material in order that the required amount of proteid matter may be obtained. By this it must not be understood that vegetable food is undesirable, for such is certainly not the case. Starchy foods are particularly valuable and a necessary part of a normal diet, the cereals especially, when properly prepared, being very completely digested and absorbed; but it is physiologically injudicious to depend entirely upon vegetable food for the necessary proteid matter. The latter is far more economically (from a physiological standpoint) obtained from animal foods, where it exists not only in a much more concentrated form, and as a consequence is more readily digestible, but the proteid matter itself is more quickly and completely assimilated than the vegetable proteid, even under equally favorable circumstances.

It is a well-understood fact, however, that the digestibility and best utilization of food depends greatly upon a reasonable variation in the character of the dietary.

[1] The Rumford Kitchen Leaflets, No. 3.

Too great sameness, especially if long continued in, may even lead to an actual impairment of the digestive organs. Hence the instinctive desire common to mankind in general for variety in the daily diet rests upon physiological grounds well worthy of recognition.

One of the most important conditions modifying the digestibility of animal proteid foods is the manner in which the muscle fibers are bound together, determining, as it does, the extent of their mechanical subdivision on reaching the stomach. Thus, in the various kinds of meat the short and delicately fibered muscles are obviously more readily digestible than the longer and tougher fibers, which plainly offer greater resistance to disintegration. It is on this account that the breast of a young chicken, for example, is so easily digestible, the short, tender fibers easily breaking apart and thus exposing more surface to the action of the digestive fluids. For a similar reason the short, flaky muscles of the more digestible varieties of fish are easily assimilated. Further, the presence or absence of fat exercises a marked modifying influence upon digestibility; as a general rule it may be said that lean meat is more readily digested, in the stomach, at least, than fat meat, although the nature of the fat present and the manner of its distribution are important factors. Thus, the presence of a hard or difficultly fusible fat, as in mutton, tends to retard the digestion of the proteid constituents of the meat more than the presence of a softer or more readily fusible fat, such as is found in beef. Again, the intimate mingling of fat with the individual muscle fibers, as in the tissues of the eel and lobster, tends to check the rate of digestibility. The same effect is seen in certain kinds of fish flesh; thus, in the shad the white meat, with its greater freedom from incorporated fat, is nearly 10 per cent more digestible than the dark and fatter meat of the same fish.

EFFECTS OF PREPARATION OF FOOD UPON DIGESTIBILITY—COOKING.

The effect of the preparation of food, and especially its cooking, upon its digestibility is a subject about which much is written and little definitely known. Here again, as in the case of the differences in digestibility of different substances, our present knowledge comes mainly from a priori considerations. The experimental data are unfortunately very meager.

The chief underlying principle is the same as that involved in the dissolving of the ore in the illustration given above. If the particles of ore are large the acid will act on them slowly. Stirring hastens the solution of the soluble materials. Time is needed for those that are slow to dissolve. If the particles of ore are incased in siliceous or other insoluble matter thorough disintegration is necessary, so that the acid may come in direct contact with the material to be dissolved. So in the digestion of food meat is cut into small pieces and chewed into still finer ones; grains of wheat which can not be well chewed are first ground. Milk requires neither cooking nor chewing. Its nutrients are either already in solution or suspension in very finely divided portions. It has no starch to be changed into sugar by ptyalin. Its only carbohydrate, milk sugar, is already dissolved; we accordingly drink it raw. Many kinds of meat are found by experiment to be digested as readily or even more readily when taken raw than when cooked, provided they are properly masticated, i. e., finely chewed; but we like their taste better, and are more inclined to masticate them properly when well cooked. The flavor induced by cooking excites the secretion of digestive juices which corresponds to the addition of acid to the dissolving ore. Furthermore, by cooking the connective tissue of some of the tougher meats is disintegrated. The collagen which serves as the binding material is changed to more or less soluble gelatin, the meat is made tender and can be chewed more finely, the further disintegration in the alimentary canal is facilitated, and the proteids and fats are more perfectly exposed to the solvent action of the digestive juices.

The following remarks of Professor Chittenden bear directly upon this subject:[1]

More important from a dietetic standpoint is the effect of cooking or preparation of proteid foods on their digestibility, and in this connection it must be remembered that digestion in the broad sense of the word is a complex process, dependent upon the harmonious working of several closely allied processes. By way of illustration it may be mentioned that a given food stuff may owe its lack of digestibility either to an inherent resistance to the solvent action of the digestive juices or more directly to the fact that its ingestion fails to cause the requisite flow of the necessary digestive secretion. Now, one of the effects of cooking is to improve, or rather develop, an agreeable flavor in the food, thus increasing its palatability and causing a pleasurable stimulation of the sense of taste. This at the same time leads to a greater outpouring of the needed digestive juices, thus furnishing the means for more rapid and complete digestion. Further, the very increase in palatability incidental to proper and intelligent cooking leads to a more thorough mastication of the proteid food, while at the same time the cooking tends to facilitate its separation and mechanical subdivision. This latter, as already stated, greatly aids the digestive process, by enabling the different digestive secretions to come into more intimate contact with the individual particles. This tendency toward disintegration and general softening incidental to cooking is due mainly to the complete or partial hydration of the connective tissue fibers by which the muscle tissue is held together into gelatin, which is far more readily digestible and assimilable than the collagen itself.

In all methods of cooking meats, as indeed all forms of proteid food, whether by boiling, roasting, or broiling, the proteid or albuminous matter undergoes more or less complete coagulation. This, however, is essential in any process of cooking by which a rich and appetizing flavor is to be developed, and if not accomplished at too high a temperature offers no obstacle to easy digestibility; indeed, it has already been stated, owing to the conversion of the collagen of the surrounding connective tissue, digestibility may be even increased. At the same time, it is well to remember that raw beef, for example, if very finely divided by chopping, is somewhat more easily dissolved by the gastric juice than when coagulated by cooking. All things considered, however, proper cooking unquestionably tends to increase the general digestibility of proteid foods. At the same time it alters the consistency and constitution of the food stuff, joining to it an odor and taste which would otherwise be wholly lacking, and, no less important, causes the destruction of disease germs or other related organisms possibly present.

The many methods of preparing proteid foods for consumption need hardly be considered here; they have little bearing on the direct digestibility of the various food stuffs, except so far as they modify the degree of coagulation of the proteid matter, the mechanical subdivision of the tissue, and the removal or addition of admixed fat. More important, however, is the indirect effect on digestibilty incidental to the degree of palatability produced, with the corresponding degree of stimulation of the secretory processes by which digestibility is so greatly augmented.

---

[1] Loc. cit.

Palatability and digestibility go hand in hand, and the intelligent preparation of a so-called cheap or tough piece of meat, for example, may result in as digestible and nutritive a product as the more careless preparation of a piece of tenderloin.

Some interesting experiments upon the rapidity of digestion of meats, cooked and uncooked, and of milk in various forms, have been conducted by Jessen at Tübingen.[1] They were made by the methods of artificial and natural digestion, those of the latter kind being performed with a dog and with a man.

To test the effect of cooking lean beef, portions were finely chopped and the tendons and other connective tissue were separated as completely as practicable. The material thus prepared was divided into several parts. One part was left raw, others were boiled, and still others were roasted. Of the boiled and roasted portions some were rare, or, as Jessen calls them, "half done," and others well done. The raw, half-done, and well-done portions were tested by artificial digestion with pepsin, by experiments in the stomach of a dog, and by experiments in the stomach of a healthy man. In the experiments by artificial digestion the meat was placed in glass tubes, an acid solution of pepsin was put upon it and the tubes with their contents were kept at about the temperature of the body for 24 hours, with occasional stirring. With the dog a fistule was used consisting of the usual metal tubes, permanently inserted through the skin into the stomach, and opened or closed with a stopper at will. The meat was inclosed in a cloth, inserted through the tube, and removed after the desired time. In the experiments with the man, a laboratory servant, the food was taken into the stomach when the latter was empty, and after digestion for the desired time, withdrawn by a stomach pump. The results of the experiments of the three kinds were essentially concordant. The raw meat was digested more readily than the cooked. In the trial by artificial digestion the residues unaltered by the pepsin were smallest with the raw meat and largest with that which had been most thoroughly cooked by boiling or roasting. In those with the man the digestion was completed in different lengths of time as set forth in the figures of Table 15, herewith. Similar trials were made with other kinds of meat and with milk. The results of the experiments with the man are summarized in Table 15.

TABLE 15.—*Relative time required for digestion of cooked and uncooked foods in a man's stomach.*

MEATS.

	Hours.
Beef:	
Raw	2
Boiled, half done	$2\frac{1}{2}$
Boiled, well done	3
Roasted, half done	3
Roasted, well done	4
Mutton, raw	2
Veal, raw	$2\frac{1}{2}$
Pork, raw	3

[1] Ztschr. Biol., 1883, p. 128.

MILK.

	Hours.
Cows' milk, 602 cubic centimeters:	
Uncooked	3½
Boiled	4
Sour	3
Cows' milk, 675 cubic centimeters, skimmed	3½
Goats' milk, 656 cubic centimeters, uncooked	3½

It will be seen that the raw veal required somewhat longer time for digestion than the mutton, and that the pork, which doubtless contained a larger percentage of fat, was still slower in digestion. The boiled milk was digested somewhat more slowly and the sour milk a little more quickly than the uncooked milk.[1]

It should be insisted that the experiments of this kind, of which the number on record is small, do not suffice for satisfactory generalizations. They can not be taken as exact measures of the degrees of digestibility of the different materials. It should be born in mind, furthermore, that they apply only to what takes place in the stomach; while the normal process of digestion goes on in the intestines after the food has left the stomach. At the same time, these results are the more worthy of confidence because they accord with the chemistry and histology of the subject.

It is a familiar fact that tough meats are made tender by long heating at a comparatively low temperature. The best explanation of this is, perhaps, to be found in the disintegration of the connective tissue with the formation of gelatin. Unquestionably the principle is an important one in its practical applications.

Vegetable foods often require cooking to fit them for use. This is especially true of starchy foods, such as wheat, corn, beans, peas, and potatoes. The starch is contained in cells, the walls of which are acted upon comparatively slowly by the digestive juices. By cooking the cells are disrupted and the starch itself undergoes more or less of a chemical change, so that it is more easily and completely acted upon by the digestive ferments.

EFFECTS OF FOOD ADJUNCTS ON DIGESTION—FLAVORING MATERIALS.

A great deal has been said, much has been written, and a small amount of accurate experimenting has been done upon the effects exerted upon the digestion of food by food adjuncts, such as spice, mustard, and other flavoring materials; beef tea and meat extract; tea, coffee, chocolate, and similar beverages, and alcoholic drinks. Professor Forster, in speaking of what the Germans call "Genussmittel"—appetizers is perhaps the nearest corresponding word in English—the materials which are taken with food either for their own agreeable flavor or to

[1] See summary of later experiments on the digestibility of milk in Experiment Station Record, 5, pp. 959, 960.

improve the flavor of the food and which are often supposed to help digestion, says in substance, as follows:[1]

There is no doubt that the human digestive apparatus can be excited to activity in various ways with Genussmittel, including such as are used by man in a refined civilization, at the beginning and end of his meals, e. g., meat broth, salt and salt condiments like caviar, cheese, etc. * * * We know that when brought into contact with the mucus membrane of the stomach and intestines of a living animal, they cause the filling of the blood vessels and secretion of the digestive juices. Sugar and salt are hardly brought into the mouth before they excite abundant effusion of saliva. Indeed, the same effect is produced even by the sight or smell of savory foods, and some of the well-tasting substances may act upon the digestive apparatus and its glands by simply being taken into the blood circulation. Thus, I have observed experimentally a rich secretion of gall after an injection of a solution of sugar into a vein (*vena mesenterica*). * * * It is very natural to infer from this that the work of digestion will go on better with the aid of such condiments than without them, in two ways: Either more nutriment might be digested from the same food, or, if there were no increase in the amount digested, it might be digested more quickly with their help, which would likewise be a gain. * * * But, important as this may seem to physiologists, it is of minor consequence with healthy persons. Thus, in experiments made with a man under my direction, when meat had been treated with water [to remove the "extractives" which give meat its flavor and which are the chief constituents of beef tea and meat extract] and was so tasteless as to be eaten in any considerable quantity with difficulty, the quantity digested and observed to pass into the circulation was as large as with the same weight of meat roasted in the ordinary way; and both Bischoff and Hofmann have found that meat extract taken with bread or with a mixed diet did not materially affect the digestion. And in experiments by Flügge with a mixed diet so tasteless as to make it, when continued some time, extremely repugnant, so that great effort was required to eat it, the digestion seemed to be unaffected thereby.

For the sick and convalescent, on the other hand, the effect of these appetizers upon the digestion is of great importance, especially where the digestive apparatus has been for a time more or less inactive and requires stimulating. Thus the observations of Kemmerich show the usefulness of bouillon and meat extract in case of enfeebled digestion.

In the case of the ore, there must be plenty of acid or it can not dissolve. If the supply of digestive juices is insufficient the food can not digest. The chief use of these food adjuncts would seem to be to stimulate the production of digestive juices. The results of later experimental inquiry and the teaching of the most reputable physiologists seem to be in line with the experiments here quoted. It would thus appear that while the materials which we call appetizers may often be very helpful where digestion is enfeebled, they are, for healthy people, unnecessary and without very great effect upon the utilization of food in the body. This subject, however, is one upon which we should speak with the greatest caution. In the present state of experimental knowledge categorical assertions are hardly justifiable. It is easy to cite statistics upon the quantities of nutrients digestible from different food materials, but all discussion of the effects of condiments, stimu-

[1] In the volume on Ernährung und Nahrungsmittel, of Pettenkofer and Ziemssen's Handbuch der Hygiene.

lants, and cooking upon ease or time of digestion—and these are the most essential things from the hygienic standpoint—must be of that general and indefinite character which is most unsatisfactory to the careful student and investigator.

As to the effect of the quantity of food upon the proportions digested, the experiments at hand seem to point to the interesting conclusion that when a moderate amount is taken it is digested more completely than a very large, or, at times, even a very small quantity; so likewise a moderate amount of water taken with the food seems favorable, while too much has been found to interfere with digestion. As to the effect of moderate exercise just after eating, observations differ, some experiments indicating that muscular labor retards digestion, others that it does not. During sleep digestion has been found to be diminished. But the conditions upon which a proper answer to such questions depends are so complex and the definite knowledge is so limited that their discussion is neither easy nor agreeable.

SUMMARY.

In considering the digestibility of food we have to take into account (1) the quantity digested, and (2) the ease and time of digestion. As regards the quantities digested from reasonable amounts of ordinary food materials by healthy people, the best experimental evidence indicates that:

(1) The protein of our ordinary meats, fish, and milk is very readily and completely digested. The protein of vegetable foods is much less completely digested than that of animal foods. Of that of potatoes and beans, for instance, a third or more may escape digestion, and thus be useless for nourishment.

(2) Much of the fats of animal food may at times fail of digestion. This is presumably true of vegetable fats, but the quantities are in general so small that the determinations of the proportions digested are not very accurate.

(3) The carbohydrates, which make up a large part of vegetable food, are in general very digestible The crude fiber or cellulose is an exception, but the quantities of this in the materials used for the food of man are too small to be of importance. Sugar is believed to be completely digested. This is assumed to be the case with the sugar of milk. The other carbohydrates of animal foods are very small in amount.

(4) The animal foods have in general the advantage of the vegetable foods in digestibility, that they contain more protein and that their protein is more digestible.

(5) The quantity digested appears to be less affected by flavor, flavoring materials, and food adjuncts, and to differ less with different persons, than is commonly supposed.

Concerning the relative ease and time of digestion of different foods, and consequent comfort and health, the lack of accurate experimental data renders it more difficult to make concise statements. Cooking and other conditions are very important. Very much depends upon the individual peculiarities of different people.

SUGGESTIONS AS TO EXPERIMENTAL WORK NOW NEEDED.

To promote a better understanding of the digestibility of different kinds of food by different individuals several lines of inquiry need to be prosecuted. They have to do with:

(1) Artificial digestion.
- (*a*) Improvement of methods.
- (*b*) Actual tests of digestibility of different materials.

(2) Actual digestion.
- (*a*) Investigation of metabolic products, their occurrence, nature, and amount.
- (*b*) Improvement of methods.
- (*c*) Systematic series of experiments on digestion.

Artificial digestion.—Concerning the experiments upon artificial digestion, it will suffice to say that what is now most pressingly needed is (1) a compilation of the results of inquiry already obtained, and (2) the elaboration of the methods of inquiry. These methods have, indeed, been worked out with no small degree of success, and it is now possible to make artificial tests of digestibility of various kinds of food materials the results of which agree tolerably well with the results of digestion tests by healthy men. But the conditions which affect digestion by the two methods are not yet well enough understood and the results by the artificial method as carried out to-day do not bring entirely accurate indications of the actual digestibility in the human organism. Of course similar remarks may be made regarding the artificial digestion of feeding stuffs and their natural digestion by domestic animals. The whole subject demands careful inquiry and promises most valuable results.

Actual digestion.—In experiments upon natural digestion by man, aside from the intrinsic difficulty of working with individual food materials, which become speedily distasteful and can not be endured by the average man for a very great length of time without danger of interruption of the normal digestive functions, the chief obstacle to the getting of accurate results is found in the metabolic products. It is greatly to be desired that investigations be made to learn more definitely what these materials are and in what quantities they are secreted under different conditions as to the characteristics of the individual and the kinds and amounts of food eaten. Reference was made above to experiments in which this subject has been studied. More inquiry is greatly needed. The investigations might be carried on some such plan as the following:

Suppose the question be, How much metabolic nitrogen will be excreted by a person of a given class, e. g., a healthy man, with a given amount of food? A ration may be planned which shall consist mostly of fats and carbohydrates and shall contain only enough nitrogenous material to make it palatable. The nitrogenous material may consist of a material, like lean beef, which has been found to be almost, if not quite, completely digestible. The quantity of undigested nitrogen in such a ration will be extremely small and its amount can be estimated approximately from the experimental inquiry already made, and still more accurately if further investigations of the digestibility of meats accumulate. From the whole nitrogen excreted by the intestines that of the undigested residue can be subtracted and the remainder taken as the measure of the metabolic nitrogen. Of course it is desirable that the nitrogenous compounds in the excreta be separated and studied as far as practicable.

In like manner the quantities of fat, or, more properly speaking, ether extract, in the metabolic products may be approximately determined. For this purpose it is desirable to devise a ration consisting mainly of protein and carbohydrates with a minimum of fats and other materials which would be extracted by ether. The nature as well as the amount of these compounds in the food can be studied. Pains can be taken to be sure that they are in as digestible form as possible. The quantity of ether extract in the undigested residue will thus be reduced to a minimum, and some information as to its nature will be had in advance. The ether extract in the excreta can then be determined quantitatively and qualitatively. Comparison of the extracts in the food and excreta will throw light upon the nature and amounts of ether extract in the metabolic products. Of course other reagents than ether might be used for the extractions. "Ether extract" is here referred to because ether is the substance most commonly used in our present method of analysis.

The above suggestions indicate in a general way the manner in which the metabolic products can be studied. Of course the individual investigator will decide upon the details. It is, however, to be hoped that some cooperation between investigators in this line may be brought about.

Meanwhile it is possible to make tolerably accurate determination of coefficients of digestibility by the methods now in vogue. The quantities of metabolic products can be roughly estimated and allowance made for them in the summing up of the results. The time is ripe for a systematic series of experiments on the digestion of our ordinary food materials—meats, fish, eggs, milk, butter, cheese, bread, crackers, and other substances made from cereals, and potatoes and other vegetable products. It would not be difficult to plan such a series of experiments and to carry them out successfully, especially if the cooperation of a number of investigators could be secured.

CHAPTER V.

PREPARATION OF FOOD—COOKING.

What has just been said regarding the effects of cooking upon digestibility covers only a small part of the ground which needs to be traversed. Among the specific questions which need investigation are the temperature best adapted to the cooking of different materials in different ways—as roasting, frying, boiling, steaming, and stewing; the chemical changes involved in the process; the loss of nutritive material in preparing it for use; the effects upon flavor, palatability, and increase or decrease of digestibility. This represents a wide and useful field of inquiry, which is almost unexplored. How little is definitely known of this subject is illustrated by the fact that in the work of König on the Chemistry of Foods and Food Adjuncts, which is by far the most complete treatise upon this subject available, and which consists of two volumes, of nearly 2,600 royal octavo pages, only nine pages are given to the cooking of food. This chapter is, however, so interesting that the substance of it is given below.

PREPARATION OF FOOD—KÖNIG'S SUMMARY OF INVESTIGATIONS.[1]

Uncivilized man takes his nourishment like animals—as it is offered by nature. Civilized man prepares his food before eating, and in ways which are, in general, the more perfect the higher his culture. The art of cooking, when not allied with a degenerate taste or with gluttony, is one of the criteria of a people's civilization.

The chief advantage of the preparation of food for eating is to facilitate digestion. This is accompanied by (1) adding condiments and giving it an agreeable appearance, so as to make it more appetizing, and thus increase the secretion of the digestive juices; (2) by loosening its texture and exposing it more fully to the solvent action which is the essential part of digestion.

EFFECTS OF CONDIMENTS UPON DIGESTION.

Flavoring materials and an agreeable appearance do not increase digestion directly, but act rather upon the nervous system, which, in turn, transmits the stimulus to the digestive organs and incites them to greater activity. While this may be helpful, it does not necessarily increase the amount actually digested from the food. Meat that has

[1] Chemie der menschlichen Nahrungs-und Genussmittel. Von Dr. König. Dritte Auflage; II. Band, pp. 138–139, 605–611, 1244–1252.

been extracted by water and made entirely tasteless has, according to Foster and Rynders, been found in actual experiments to be as quickly and completely digested as an equal weight of meat roasted in the ordinary way; and the same is stated by Flügge to have been the case with a mixed diet so tasteless as to be repulsive. The principal condiments that influence digestive secretions are salt, sugar, alcohol, and spices.

Salt increases the flow of saliva and of gastric juice, hence one value of its use at meals in side dishes like caviar, olives, and pickles. The salting of food is so indispensable that salt is in some regions considered a necessity, and the lack of it has been a cause of war.

Sugar is very agreeable to the palate, and the mere thought of it often makes the mouth water; in other words it causes the secretion of saliva. Doubtless the flavor is the chief cause of the great demand for sugar, for although it is a valuable aliment, it hardly surpasses starch, dextrin, and other carbohydrates in nutritive value.

Alcohol.—Taken in moderate quantities in such forms as cognac, brandy, wine, beer, and other beverages, alcohol is likewise an important stimulant to digestion. Brandy, whisky, sherry, and the like are therefore favorite remedies in disturbances of the bowels and stomach, and this helps to explain why the poorer classes, who often live upon a wretched diet of the less digestible foods, such as coarse bread and potatoes, have a craving for strong and stimulating alcoholic drinks. It is the improper and excessive use of alcoholic beverages which makes them a scourge to man by weakening his digestive apparatus and undermining his general health.

At the same time it should be said that sugar and alcohol, until they are absorbed by the system, hinder, temporarily, the actual process of digestion, at least in the stomach. After their disappearance from there the digestion goes on more vigorously than if they had not been taken.

The influence of carbonated waters and salts is often very beneficial.

Spices.—Some of these act directly upon the glands which secrete the digestive juices, as is the case with pepper, which contains piperin, and with mustard, which contains an oil that increases the flow of bile. Others first delight the sense of smell, which then stimulates the flow of saliva. Such are vanilla, cinnamon, and cloves. A similar influence is exerted by onions, parsley, and the like, which contain aromatic substances. Fruits, likewise, contain, besides aromatic oils, more or less malic acid, which directly aids digestion.

Meat extract, tea, coffee.—These are stimulants, and, like wine and beer, excite more particularly the nervous system, and in so doing they are at times very useful. The stimulating agents in tea and coffee are alkaloids and essential oils; those in beer and wine are alcohol and fragrant ethers. Tea and coffee, taken in moderation, have no disturbing effect upon digestion.

Cooking changes the texture of food, making it in some cases more and in others less digestible. In general, the digestibility of vegetable materials, as flour and potatoes, is increased, and that of animal foods, as meat, is diminished by cooking, though there are exceptions. Cooking can not add to the amount of nutritive material in food, except in so far as it may increase the proportion which is actually digested; but it may, and often does, remove considerable quantities, as in the boiling of meats.

We may consider the effects of cooking under different categories, as the boiling, broiling, and roasting of meats, the boiling and baking of vegetables, and the baking of bread.

BOILING, BROILING, AND ROASTING OF MEATS AND OTHER ANIMAL FOODS.

Boiling, as commonly understood, is heating in water. This is done generally with the water at or near the boiling point. But much cooking in water is done at a lower temperature, because the water is not always heated to the boiling point, and, more especially, because when meats and other materials are immersed in boiling water, as is usual in cooking, the heat does not penetrate so as to bring the temperature of the interior up to that of the water. On the other hand, there are cases where, as in the preparation of gelatin, the water and the meat or bone or other material are heated much above the temperature of boiling water.

Steaming is similar to boiling, the difference being that the material is surrounded by steam rather than water.

Boiling and steaming effect four kinds of change in the food. (1) It is made more soft and tender. In meats the connective tissue is softened, weakened, and, to a greater or less extent, gelatinized and dissolved, and thus the fibers of the meat are loosened. (2) Some material is made soluble. This is the case with the gelatinoid substance of meat as just stated, and especially with that of bone. By long boiling the albuminoid materials of meat are dissolved to slight extent. On the other hand, more or less of the albuminoids of meat, as the albumin and myosin, are coagulated and hardened by boiling. (3) Flavors are developed which make the materials, especially meats and fish, more agreeable to the taste, and hence tend to promote the secretion of digestive juices and thus aid digestion. (4) More or less material is dissolved out of the food by the water. When used for soup the dissolved materials are well utilized, otherwise they are lost.

Various experiments[1] have proved that boiled, roasted, or smoked meat, for example, is not as quickly and completely digested as raw meat. Our preference, then, for meat so prepared only shows how

[1] Chittenden and Cummins, Am. Chem. Jour., 6, 318; M. Popoff, Ztschr. physiol. Chem., 1890, 14, 524; A. Stutzer, Centbl. allgem. Gesundheitspflege, 1892, 59; Jessen, see above, p. 75.

much value we attach to the pleasing tastes, odor, and physical condition of our food.

The cooking of meats and fish in "dry heat," by roasting, broiling, or frying, likewise effects several kinds of change. (1) The albuminoids are coagulated. The result of this generally, though not always, is to harden the fiber more or less and thus make the flesh less easily and completely digestible, especially on the outside. (2) Flavors are developed which make the flesh much more palatable and tend to favor secretion of the digestive juices. Thus, while the heating tends to make the material less digestible, the effect may perhaps be counteracted in part by the providing of more of the ferments which digest the food. (3) More or less of the juices of the meat are driven out by the heating. These commonly are saved in roasting, but lost to a greater or less extent in boiling.

When meats and fish are heated, whether in roasting, broiling, frying, or even boiling, much water is driven off.

Boiling of meat.—Meat contains from 5 to 8 per cent of substances soluble in water, namely, albumen, meat bases (creatin, creatinin, sarkin, etc.), organic acids, glycogen, inosit, and salts. When meat is boiled with water albumen is rendered insoluble and remains in the meat tissue, or produces the scum floating upon the broth. Some connective tissue is transformed into gelatin and is dissolved; a part of the melted fat passes into the broth.

In boiling meat two methods are employed; either (1) the meat is placed in cold water, which is then heated more or less slowly to boiling and kept boiling for some time, or (2) the meat is put into water already boiling. The results differ. In the first case the cold water penetrates the piece of meat and extracts more or less of the juices and soluble constituents; upon boiling the albumen partly collects upon the surface of the liquid as scum. In the second case only a little of the soluble material is extracted. The albuminoids at the surface of the meat coagulate and form an impenetrable layer. This cooking prevents the extraction of the soluble materials, and the inner parts are left more or less juicy. Consequently, to obtain a strong, nutritious broth, consommé, or soup, we use the first method, warming the water slowly, but if the boiled meat is to remain juicy and is to be eaten as meat, the second method is followed. This view is confirmed by experiments of A. Vogel,[1] who found that meat cooked by gradual warming with cold water lost much nitrogen and gave a richly nitrogenous broth, while meat placed directly in boiling water retained its nitrogen and juice but yielded a poor broth.

As a general thing, however, meat is not completely extracted even in making soup, and bones are often boiled with it to secure a strong broth.

Von Wolffhügel and Hüppe[2] have observed the temperature in the

[1] Chem. Centbl., 1884, 639.

[2] Mittheilungen des kaiserlichen Gesundheitsamts, I, and Abst. in Jahresber. Tier. Chem., II, 1881, 441.

interior of cooking meat and found it always considerably lower than the outer temperature. In a piece of meat weighing 4.5 kilograms the interior temperature after four hours' boiling was only 88° C. The inner temperature of meat which was being roasted varied from 70 to 95° C., according to the size of the piece.

When canned meat in large and even in small cans was kept in a salt-water bath at 102 to 109° C., the interior temperature of the meat rose only to 72 to 98° C., according to the size of the cans. This explains why large cans of meat always have more bad spots than smaller ones. The heat is not sufficient in the large cans to destroy the bacteria or other organisms that cause the meat to decompose. It may be that the formation of an insoluble layer of albumen prevents the penetration of both the boiling water and the heat.

Meat broth.—The quantities of ingredients in meat broth are illustrated by an experiment. We boiled 500 grams of beef and 189 grams of veal bones in the ordinary (German) household way, and obtained a little more than half a liter (543 cubic centimeters) of strong broth or soup. This contained by weight:

	Per cent.
Water	95.18
Total dry substance	4.82
Nitrogen	.19
Protein (N. × 6.25)	1.19
Fat	1.48
Extractives (carbohydrates, etc.)	1.83
Ash	.32
Potash	.15
Phosphoric acid	.09

Good broths may also be made from less meat and more water, by adding savory herbs to improve the flavors. A. Payen[1] gives the following figures for the materials used in making broths and the composition of the broths as actually prepared from them:

A.—*Materials used for making meat broths.*

	Meat.	Bones.	Salt.	Vegetables and spices.	Water.
	Grams.	*Grams.*	*Grams.*	*Grams.*	*Grams.*
No. 1	500				1,500
No. 2	1,433.5	430	40.5		5,000
No. 3	500		8	32.2	2,000

B.—*Percentage composition of meat broths prepared from above.*

	Water.	Total dry residue.	Organic substance.	Salt.
	Per cent.	*Per cent.*	*Per cent.*	*Per cent.*
No. 1	98.41	1.59	1.27	0.32
No. 2	97.21	2.79	1.68	1.11
No. 3	97.95	2.05	1.25	.80

[1] Substances Alimentaires, 1865, 94–105.

It appears from these analyses that the amount of solids in meat broths is generally small. Consequently their strong taste and stimulating effect upon the nervous system must be ascribed to the meat bases, creatin, carnin, etc., and potassium salts.

A bowl of nourishing soup or bouillon made in the ordinary way—from 3 to 4 grams of meat extract with the addition of spices, egg, salt, etc.—contains only 4 grams of solids. Of these, 3.4 grams is organic matter, 0.34 gram nitrogen, and 0.8 gram is salts having about 0.3 gram of potash.

Besides meat bases, soups contain also more or less gelatin, varying directly with the quantity of bones used in the making. From the composition of bones such as occur in ordinary meat it may be estimated that 100 grams would yield by ordinary kitchen boiling:

	Grams.
Total dry substance	2.0 to 7.5
Nitrogenous matter	0.2 to 2.8
Fat	0.6 to 5.5
Other organic matter	0.1 to 0.5
Salts	0.1 to 0.2

After the meat has been boiled to make broth it still contains meat fiber, part of the albumen, connective tissue, and fat, and if not completely extracted, small residue of meat bases also. Boiling disintegrates the fibers. This is the reason why boiled meat can be more easily chewed and is preferred to raw meat.

Boiled meat.—Of course boiled meat has not the same nutritive value as fresh raw meat. Aside from the fact that meat may be rendered less digestible by boiling, it also loses nutritive material. Dogs have died when fed upon meat which has been extracted by boiling. As it is chiefly the withdrawal of salt which makes the extracted meat an incomplete food, the restoring of the salts will increase its nutritive value and lessen its injurious effects. Meat which has been long boiled should not be used as exclusive diet without at least the soup which it has served to make. It should be remembered, however, that when the meat is put into boiling water at the outset, especially if it is in large pieces, it loses comparatively little nutritive material.

Roasting, frying or broiling is a decidedly more rational way of cooking meat, since the juices are mostly, if not entirely, saved and the tissue and fiber are at the same time loosened by the heat, or by the steam and the fat vapors which the heat generates. In roasting and frying meat a crust is formed on the outside by the coagulating and hardening of the albuminoids. It is supposed that at the same time trifling quantities of carbon and nitrogen are driven off and a small amount of acetic acid is produced which dissolves some of the ingredients of the meat. The fat undergoes a partial decomposition into fatty acids and glycerin, and a little of it is volatilized.

The following analyses, made in König's laboratory, show the comparative composition of specimens of fresh meats, boiled and roasted.

Comparative composition of meats before and after cooking.

	Water.	Nitrogenous matter.	Fat.	Extractive matter.	Salts.
Beef:	*Per cent.*	*Per cent.*	*Per cent.*	*Per cent.*	*Per cent.*
Before cooking (raw)	70.88	22.51	4.52	0.86	1.23
Same after boiling	56.82	34.13	7.50	.40	1.15
Same after broiling (as beefsteak)	55.39	34.23	8.21	.72	1.45
Veal cutlets:					
Before roasting (raw)	71.55	20.24	6.38	.68	1.15
Same after roasting	57.59	29	11.95	.03	1.43

The water decreased by roasting and boiling from about 71 to 57 per cent. But such analyses as these permit of no accurate conclusions as to the other changes and losses by roasting. One trouble with them is the difficulty in obtaining perfectly homogenous pieces of meat with the same proportions of fat for the comparative tests. Furthermore, raw prepared meat dishes can not be compared exactly in composition with the same meats after they have been roasted or boiled or fried, because the fat added in the ordinary cooking of beefsteak, cutlets, or roasts alters the composition of the cooked product. On the other hand, cooking without fat would be abnormal, since precisely this addition of fat hinders the decomposing and volatilizing of other constituents of the meat. These considerations should be taken into account in comparing the figures of the table herewith in which the figures of the preceding table are recalculated to the basis of dry matter; that is to say, the water is taken out and the composition of the water-free substance is shown in each case.

Comparative composition of water-free substance of meats before and after cooking.

	Nitrogen.	Nitrogenous matter.	Fat.	Extractive matter.	Salts.
Beef:	*Per cent.*	*Per cent.*	*Per cent.*	*Per cent.*	*Per cent.*
Before cooking	12.37	77.31	15.47	2.98	4.24
After boiling	12.65	79.06	17.38	.90	2.66
After roasting	12.27	76.73	18.41	1.59	3.27
Veal cutlets:					
Before cooking	11.39	71.17	22.45	2.32	4.06
After roasting	10.93	68.36	28.18	.09	3.37

These figures show, notwithstanding the objections just mentioned, that in roasting, as in boiling, some extractive matter and some salts are removed from the meat. The reason is that in the roasting, as everyone knows, a small part of the juice oozes out. To take the latter as gravy with the roast is, therefore, entirely rational.

Boiling of bones.—When bones are boiled more or less of their nitrogenous cartilage (ossein) is converted into soluble gelatin. If the boiling is done under high pressure nearly all cartilage is removed from the bones, and after certain manipulations comes into commerce in thin, brittle, colorless tablets. This pure gelatin is used for making jellies, puddings, etc. Glue is impure gelatin. Soft and spongy bones yield most gelatin. Large, hollow ones are broken into small pieces before extraction.

Boiling of milk.—The object of this is to kill the bacteria in the milk and thus for a time prevent its souring. In the boiling a film forms on its surface, which consists of casein, and the peculiar odor given off is due to sulphureted hydrogen.[1] Apparently no other important changes in composition take place; but E. Jessen[2] finds, nevertheless, that boiled milk is less rapidly digested than raw milk.

Boiling and baking of vegetables.—Vegetable foods are more difficult to digest than animal foods. Their preparation is therefore more complicated and thorough. The nutritive substances are inclosed in cells, often with thick walls, and are not readily acted upon by the digestive fluids. But when vegetables are boiled the cell contents expand and burst through the walls. The fragrant and savory substances are set free with the other materials which have been inclosed in the cells, and their astringency and bitterness are tempered. Some of the constituents are dissolved by the water or suffer other change. Starch, an important ingredient of many vegetable foods, such as potatoes, wheat, rice, and oat meal, takes up water and assumes the soft, pasty condition which is necessary for its transformation into soluble dextrin and sugar. The boiling of foods, especially vegetables, may therefore be called a preparatory digestive process.

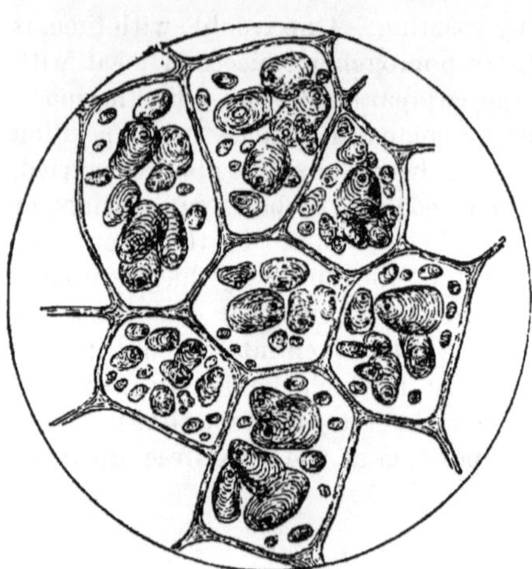

FIG. 2.—Cells of a raw potato, with starch grains in natural condition.

The accompanying diagrams[3] (figs. 2, 3, and 4) illustrate the differences in the structure of the starch grains of the potato. No. 1 shows the cell in the raw state. In No. 2 the potato had been boiled a half hour. The starch granules are swollen, but not enough to entirely fill the cells. In No. 3 the potato had been well steamed and then mashed; the cells are filled to distention and a number have burst open so that the starch, now an entirely amorphous mass, is considerably scattered.

Cells from other plants behave like those of the starch-bearing tubers of the potato. The seeds of legumes, such as beans, peas, and

---

[1] Schreiner, Amtl. Bericht. d. 50. Versammlung deutscher Naturforscher und Aerzte, München, 1877, 218. Abst., Jahresber. agr. Chem., 20, 1877, 422–423.

[2] See above, p. 75.

[3] Märcker, Studien in der Spiritusfabrication.

lentils, are, in their natural state, difficult to digest, because their starch granules lie closely packed within the indigestible cell walls. On boiling the starch swells, the cells burst open, their contents are changed into a pulpy mass which, when strained through a fine sieve, makes a very nutritious and digestible dish. The predominant constituent of the nitrogenous matter "legumin" is dissolved and made more digestible by the phosphate present; but it forms an insoluble compound with lime; hence these vegetables can not be boiled advantageously with hard water.

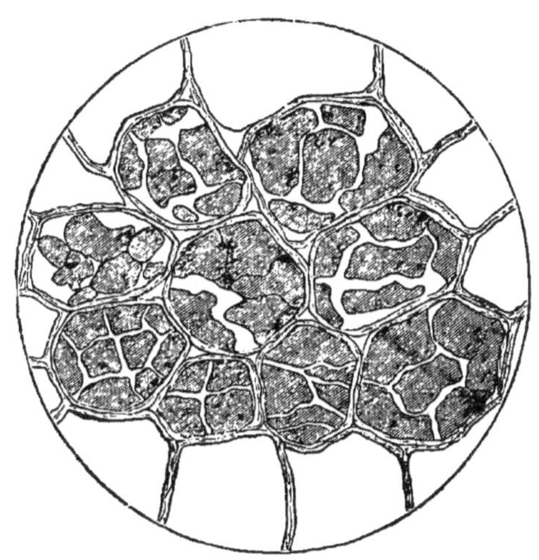

FIG. 3.—Cells of a potato boiled in water one-half hour.

Flour and starch in their ordinary dry state are hardly fit for food for man. Cooking with milk or water softens and gelatinizes them sufficiently for making puddings, cakes, omelets, pies, macaroni, etc., but for cakes or bread a porous light dough is obtained by particular treatment with yeast or baking powder, as will be explained beyond.

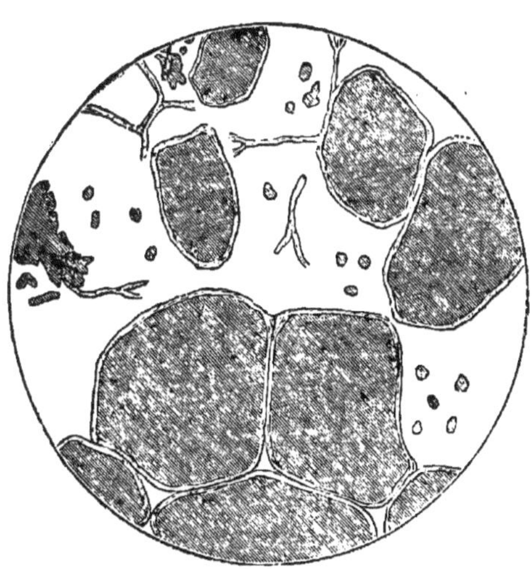

FIG. 4.—Cells of a potato well steamed and mashed.

By the baking and roasting of starchy foods some starch is also converted into more assimilable dextrin, as in the hard, brown crust of bread, the surface of toasted bread, baked potatoes, and macaroni.

Soups and soup making.—The following tables illustrate the composition and nutritive values of some home-made soups. The data are from experiments in König's laboratory. The soups were made in accordance with the methods practiced in ordinary (German) households, to wit: Pea soup, by boiling peas with smoked sausage and straining off the pulp; potato soup, by boiling potatoes with waste pork; bread soup, from pieces of bread, sugar, and water; farina soup, from barley farina and milk.

Composition of soups (as made in German households).

	Specific gravity.	Water.	Nitrogenous matter.	Fat.	Extractive matter (carbohydrates, etc).	Cellulose.	Ash.
		Per cent.	*Per cent.*	*Per cent.*	*Per cent.*	*Per cent.*	*Per cent.*
Pea soup	1.0540	88.26	3.38	0.93	5.60	0.70	1.13
Potato soup	1.0385	90.96	1.37	1.55	4.87	.26	.99
Bread soup	1.0455	88.81	1.25	0.16	8.92	.38	.48
Farina soup	1.0415	87.66	2.44	1.48	7.46	.09	.87

Evidently these figures can have no general value, as the composition of soups must vary with the materials used and the methods of preparation, neither of which are ever uniform.

Loss of nutritive substance in the boiling of vegetables.—This is illustrated by some observations by P. Wagner and K. Schaefer.[1] The following are their figures:

Loss of mineral matters in the boiling of potatoes with and without skins.

	The boiling water poured off from 1 kilogram of potatoes contained—			Lost from original amount in the potatoes.		
	Total mineral matter.	Potash.	Phosphoric acid.	Total mineral matter.	Potash.	Phosphoric acid.
Unpeeled potatoes:	*Grams.*	*Grams.*	*Grams.*	*Per cent.*	*Per cent.*	*Per cent.*
Boiled	0.28	0.10	0.02	3.64	3.32	1.12
Steamed	.09	.03	.005	1.17	.69	.03
Peeled potatoes:						
Boiled	2.15	1.25	.35	28.86	38.33	22.87
Steamed	.55	.26	.07	7.28	6.93	4.57

After the boiling of other vegetables as well as potatoes the excess of water is commonly thrown away. That this may involve considerable waste is shown by the following experimental results:

Ingredients in waste water from 1 kilogram of green vegetables.

	Total solids.	Nitrogenous matters.	Nitrogen-free extractives (carbohydrates, etc.).	Total inorganic matter.	Potash.	Phosphoric acid.
	Grams.	*Grams.*	*Grams.*	*Grams.*	*Grams.*	*Grams.*
Spinach	8.578	1.084	3.519	3.375	2.326	0.322
Carrot tops (chopped)	15.252	3.312	5.609	6.331	4.196	.348

The loss varied from 9 to 18 per cent of the total amount of soluble matter in the unboiled vegetable foods.

[1] Sächs. landw. Ztg., 1885, 33, p. 369. Abst., Jahresber. agr. Chem., 28, 1885, 443.

In the making of flour from cereal grains the less digestible outer coating is removed by the process of milling, and the more easily and completely digestible material is left in the flour or meal.

The starch grains of the product thus prepared are inclosed in cells, the walls of which, as explained above, make it difficult for the digestive juices in the alimentary canal of man to get at them and prepare them for absorption. We accordingly mix the material with water or milk and heat it. The starch swells, the cells burst open, and a paste or dough is formed, and thus the material is partly prepared for digestion. But in the baking of bread still further steps are taken which result in loosening the mass or making it "light," to use a common phrase, while at the same time the taste is greatly improved.

Flour simply mixed with water and baked is neither as palatable nor as digestible as is to be desired. It is difficult to chew, and the small, impenetrable pieces when swallowed resist the action of the digestive fluids. Hence, in the making of the dough some substance or substances are added to make it "rise"—that is, to generate in it a multitude of small bubbles of gas. Such materials are leaven, yeast, and baking powders. The two former convert some starch of the flour by fermentation into alcohol and carbonic acid; the latter, which usually contain mixtures of suitable chemicals, generate carbonic acid upon contact with the moisture of the dough. In making so-called "aerated bread" carbonic acid ready made is worked into the dough by machinery in closed vessels. When dough thus prepared is put into the hot oven the bubbles of gas which permeate it are expanded and thus make larger cavities. At the same time the minute, hard granules of starch, hitherto unchanged, are by the moist heat softened and gelatinized. The ferments of yeast and leaven are killed by the heat and thus their further action upon the bread is prevented. The outer surface of the bread is more heated than the interior and much of the starch is changed to dextrin, which is more soluble and believed to be more digestible. Hence the crust is often recommended for persons with enfeebled digestion. The same change occurs in the outer, browned surface of pieces of bread when toasted.

EFFECTS OF COOKING AS STATED BY HALLIBURTON.

To the statements of Professor König, above cited, the following summary of the effects of the cooking of food, by Professor Halliburton, of King's College, London, may be added:[1]

> The cooking of foods is a development of civilization, and much relating to this subject is a matter of education and taste rather than of physiological necessity. Cooking, however, serves many useful ends:
>
> (1) It destroys all parasites and danger of infection. This relates not only to bacterial growths, but also to larger parasites, such as tapeworms and trichina.
>
> (2) In the case of vegetable foods it breaks up the starch grains, bursting the cellulose and allowing the digestive juices to come into contact with the granulose.

---

[1] The Essentials of Chemical Physiology, p. 34.

(3) In the case of animal foods it converts the insoluble collagen of the universally distributed connective tissues into the soluble gelatin. By thus loosening the binding material, the more important elements of the food, such as muscular fibers, are rendered accessible to the gastric and other juices. Meat before it is cooked is generally kept a certain length of time to allow *rigor mortis* to pass off.

Of the two chief methods of cooking, roasting and boiling, the former is the more economical, as by its means the meat is first surrounded with a coat of coagulated proteid on its exterior, which keeps in the juices to a great extent, letting little else escape but the dripping (fat); whereas in boiling, unless both bouillon and bouilli are used, there is considerable waste. Cooking, especially boiling, renders the proteids more insoluble than they are in the raw state; but this is counterbalanced by the other advantages that cooking possesses.

ATKINSON'S DEFINITION OF COOKING.

The subject is viewed from a somewhat different standpoint in the concise definition by Dr. Edward Atkinson:[1]

We may define the art of cooking as consisting in applying heat to each of these subjects in such a way as (1) to render it digestible, so that its nutrient properties may be assimilated in true proportion in the human system; (2) to render it appetizing by the development of its own specific flavor; (3) to combine different kinds of food material in such a way that each will render the other palatable; (4) to remove certain portions which may not be palatable or digestible after the first application of heat, either as waste, like bone, as excess, like much of the fat that may be used for other purposes, or as woody fiber in many vegetables; (5) to add to the essential elements salt in its due proportion in almost every process, and sugar in some combinations, and other condiments, spices, or flavorings in such a way as to develop rather than to disguise the true flavor of the principal food material entering into each dish.

INVESTIGATIONS BY MRS. RICHARDS AND MRS. ABEL.

In 1889 and 1890 a grant was made by the trustees of the Elizabeth Thompson fund for experiments upon cooking. This was supplemented by private gifts for the same purpose. The experiments were carried out by Mrs. Ellen H. Richards and Mrs. Mary H. Abel. They included some trials with the Aladdin oven, the invention of Dr. Edward Atkinson, through whose instrumentality the investigations were undertaken. The following statements are from the reports of Mrs. Richards and of Mrs. Abel to the trustees of the Elizabeth Thompson fund. They are taken from a paper by Dr. Atkinson on "The right application of heat to the conversion of food material," read at the meeting of the American Association for the Advancement of Science, in Indianapolis in August, 1890:

The investigation of certain scientific principles of hygienic and economic cookery, in aid of which a grant of $300 was made from the Elizabeth Thompson fund, was to cover four points:

(1) The determination of the time and temperature required for the best results in flavor and digestibility of the several classes of food materials, i. e., meat, vegetables, etc.

---

[1] Suggestions Regarding the Cooking of Food, published by the U. S. Department of Agriculture, 1894.

(2) To ascertain the difference, if any existed, in the chemical character of bread quickly baked and that baked for a long time.

(3) To endeavor to secure a simple and effective measure of oven heats.

(4) To modify the ovens in common use in order to obtain more digestible food and to secure more economical use of food materials.

The study of these points was made by Mrs. Mary Hinman Abel in connection with the work of the New England Kitchen, thus securing more than mere laboratory results. The analyses were made in the laboratory of sanitary chemistry of the Massachusetts Institute of Technology by Messrs. G. L. Heath, F. S. Hollis, W. R. Whitney, and Misses Bragg, Day, Sherman, and White. The apparatus is being used constantly in further experiments.

The accompanying report gives in detail the results of the six months' work. It may be briefly summarized as follows:

(1) Two standard dishes have been perfected which have stood the test of six months' daily sale with constantly increasing popularity. The time and temperature necessary for these dishes are known, as well as the exact proportions of the necessary ingredients. Meat and vegetables are represented. The two dishes are beef broth for invalids, and pea soup. From the beef broth other soups are made.

(2) The change in the composition of bread when baked a long time at a moderate temperature has been demonstrated.

(3) The question of an oven thermometer has not been satisfactorily settled.

(4) In regard to the American cook stove we have come to the conclusion that no general improvement in the character of prepared food can be expected with so crude and unreliable a means. As we have demonstrated that for most substances a long time and a moderate heat is required for the best results, we must find an apparatus to secure this. Nothing which we have seen or heard of quite meets the ideal requirement.

In our practical experiments we have confined ourselves to the effect of a moderate degree of heat, continued for a considerable time. We have not considered to any extent the methods of broiling or so-called roasting, because it is well known that only tender, high-priced cuts of meat can be thus made fit for the table.

ESSENTIALS FOR GOOD COOKING APPARATUS.

For ordinary cooking apparatus the following are essential points:

(1) The degree of heat should be under perfect control, increased, diminished, or withdrawn at will, and without loss of time. This can only be perfectly attained with liquid or gaseous fuel. Solid fuel demands constant and equable running, and gives best results in large masses. The small fire box of a cook stove, and the urging of the fire for a short time three times a day are fatal objections to the use of anthracite.

(2) A tightly closed vessel heated by steam, or hot water, or hot air, offers many advantages over the top of a red-hot stove or the inside of a nearly red-hot cast-iron oven for cooking, except for the broiling and the roasting of meat and for some other methods of cookery which require the quick application of heat.

(3) For all purposes of slow cooking the oven should have a non-conducting covering which retains the heat where it is wanted, and also allows of tight closing and of security from the constant watching required by the fitful heat of a stove.

This use of a close oven with a non-evaporative atmosphere, seems to be the secret of the retention of the delicate and volatile flavors which usually flavor the house and street, and not the food as it is brought to the table.

Three kinds of apparatus are now in the market which meet more or less of these requirements, and all may be used with a kerosene lamp. The Arnold steam cooker uses steam generated in a sort of flat boiler. For some purposes, such as the cooking of cereals, as mush, and also for the prepartion of a few quarts of soup, and for the slow cooking of meat in its own juices, we find this cooker very effective. The liability to leakage and the difficulty of mending are drawbacks, and there is no non-conducting covering.

The Wanzer cooker we have not been able to purchase, as it is held by an English patent. It has a special lamp of some merit and it has good points, as the demonstration which the patentees gave us showed. The loss of heat is, however, very large and the lamp will run only four hours.

The Aladdin oven or covered stove. This is a square or oblong box of sheet iron of any desired size, with a non-conducting covering of magnesian cement or wood pulp, and is heated with a kerosene lamp or gas burner. The size in use for these experiments is 18 by 12 by 14 inches, and gives a cooking space at least equal to that of a No. 8 Crawford cook stove, and when empty can be heated to about 300° F. in an hour, and maintain that temperature for 8 hours by a single kerosene lamp of the Rochester-burner type, with the consumption of 1 quart of kerosene. When well filled with food materials in small portions the heat is sufficient to heat them in about twice the time allowed by an ordinary cook stove. When the space is completely filled with a vessel containing, for instance, 40 pounds of meat and bone and 15 quarts of water, the whole is raised from a temperature of 70 to 180° F. in 7 hours, and to 212° in 12 hours. If the lamp is then taken away or allowed to go out the temperature does not fall below 190° for 4 hours.

For this 12 hours, 1½ quarts of kerosene are needed, or a gas burner can be used. For simplicity, effective use of heat, economy of fuel, and development of flavor in the food cooked, combined with increase of its digestibility, the Aladdin oven is an apparatus far exceeding in merit any other now in market. It will not meet all the demands that the modern cook now makes of the kitchen stove, and it may be in several respects improved, but in the application of well-known and long-tried scientific principles to the cookery of food, it is a distinct advance and a most valuable invention. * * *

COOKERY OF MEAT.

The ideal preparation of a food for human use requires that the nutriment it contains should be utilized to the fullest extent, and this implies not only that it shall be in such a state that the digestive juices can best act on it, but that these digestive juices shall be properly stimulated to do their work by the taste or flavor of the food. Therefore in the cooking of meat we undertook to answer this question: What method can be employed that will yield the most in nutrition and flavor? It was determined to experiment first with beef of the cheapest cuts, as the neck and shin, these cuts although tough being among the richest in nutriment as shown by analysis, and to utilize this nutriment in the form of broth or as a basis for soups.

In examining a shin of beef we find it to consist, as far as our uses are concerned, of, first, muscular fiber; second, connective tissue; third, bone. We have here three distinct food materials, which if treated separately would require quite different processes.

(1) Muscle fiber. To prepare this for the digestive juices requires only that slight application of heat that develops the flavor which is most agreeable to the civilized palate, and so increases the nutritive value of cooked muscle fiber over raw, at least for the civilized stomach. A familiar example of the most perfect method of cooking the muscle fiber alone is broiling.

(2) The connective tissue and tendons. We find these intimately connected with the muscle fiber, and their food material is finally obtained mostly in the form of gelatin. To render these substances available for food they must be first hydrated, and then to a greater or less extent dissolved. The length and difficulty of this process differs with the age of the animal, its food, and also on the length of time the meat has been kept after killing, all of which affects the toughness of the enveloping membrane. The connective tissue in a sirloin steak is so tender that the heat necessary to cook the muscle fiber, as in broiling, is sufficient also to hydrate the connective tissue by merely heating the water contained in the steak; that of the tougher cuts, like that of the shin, need a much longer application of heat.

(3) The bones also contain a substance which yields gelatin, and to extract it

requires a long application of heat. It became evident at this step in the investigation that since muscle fiber, connective tissue, and bone must be cooked by the same process, and that this process must be a long one in order to affect the necessary changes in the connective tissue and the bone, some means must be devised to prevent the overcooking of the muscle fiber from dissipating its flavor. This settled the first requirement of the cooking vessel to be used, namely, that it must be tightly closed. On this account the ordinary iron pot was rejected, but the long famous Papin soup digester with its tightly screwed top was given a good trial on the kitchen range. But other requirements were to be met. The perfect hydration of the gelatin yielding connective tissue of meat and the proper cooking of the albumen would require that the temperature be perfectly under control. It was found to be impossible to regulate the temperature inside the digester; for, added to the ordinary difficulty found in using the variable kitchen range, was the fact that the tightly closed digester gave no sign of the rising temperature till the mischief was done. Placed over the more easily regulated flame of gas or kerosene so large a vessel showed great unevenness in the temperature in the top and bottom, and both of these methods were expensive as to fuel.

We next tried cooking the meat in earthen jars placed in the Aladdin oven to which the heat of a kerosene lamp was applied, and the results obtained were so good that it became evident that we were on the right track. The meat and bone placed in a jar and covered with cold water rose slowly and steadily, without any attention on the part of the cook, to a required temperature and could be held there for any length of time by simply lowering the flame of the lamp. The non-conducting shell of the oven assured nearly the same heat to the bottom, top, and sides of the cooking vessel, and did the work with a minimum amount of fuel.

We had aimed to produce a food that should hold nitrogen compounds in solution with all the flavor available. This would require that the extraction of the food material from the meat should be effected as nearly as possible before that temperature was reached at which the flavor once developed would be dissipated by the escaping steam. It was found by later experiment that the agreeable flavors peculiar to boiling soup were not brought out till the boiling point was nearly reached. This by our present method is 12 hours after the beginning of the cooking, and near the end of the process, so that the least possible quantity of these flavoring substances is lost by further cooking.

We, therefore, had a tin-lined copper vessel holding 30 quarts made to fit the oven, thus utilizing the entire inside space. Three of them have been in constant use in the kitchen since our early experiments demonstrated their value. By this method only have we been able to meet our requirements for the proper cooking of meat of the tougher cuts, and at the same time extract from bone and tendon a due amount of gelatin. We consider it quite probable that a steam apparatus might be devised that would do the work on exactly the same principle by surrounding the cooker with a heated medium easily regulated and using low pressure; the steam jacket of the restaurant does not answer the requirements, but to have something of this kind constructed did not come within our means.

Whether the Aladdin oven is the ultimate best form of a cooker we do not attempt to say, but it certainly deserves a high place because constructed on the principle of holding the heat to its work by a non-conducting covering, and for using an easily regulated fuel. Other contrivances examined by us, however convenient and ingenious, are on the old lines and show no distinct advance toward an application of scientific principles to cooking methods.

The method employed in the Aladdin oven is the same whether the meat and bone are cooked with a small quantity of water and used in the form of a stew, or with a larger quantity of water, the meat at the end of the process being pressed dry of its juices. This latter method is the one most employed at the kitchen because of our large consumption of this broth, merely salted as beef broth for invalids, and as a basis for our various soups.

COMPOSITION OF BEEF JUICE, BEEF TEA, ETC.

In view of the unexpected demand for the broth for the sick room we were obliged to study the composition of the various preparations in use and the possibilities of the yield of meat under various kinds of treatment.

Beef juice obtained from the best steak which has been merely warmed through over the coals and then entirely deprived of soluble substance by a screw press, is undoubtedly the most concentrated of the liquid foods. If prepared with the most scrupulous care, from the best material, and used at once, it probably leaves nothing to be desired. But in unskilled hands the risks are considerable in using this raw and most easily putrescible material. It is also a slow, laborious, and expensive operation.

Beef tea, as ordinarily made, is of uncertain composition; it may be only the juice of the meat set free by the coagulation and shrinking of the fiber on heating. Such is the beef extract made by heating chopped steak in a bottle. It may be an aqueous infusion of very variable strength, containing chiefly phosphates, kreatin, and certain extractive matters, agreeable to the palate, but of little nutritive value. It occurred to us to prepare on a large scale a broth of constant composition from both meat and bone, in such a manner as to secure a nutritive value at least equal to that of milk (without its fat), and without sacrificing the appetizing flavor. The bone gives a proportion of gelatin which, when flavored with the meat extract, is believed to be of high nutritive value.

The following table gives a comparison of these different preparations:

	Meat.	Total solids.	Solids, juice filtered before coagulation.	Solids, juice filtered after coagulation.	Coagulable albumen.	Extract of meat.	Salts or ash.
Beef juice from meat slightly broiled and pressed (round)....	*Per cent.* 26.8	11.9	10.8	4.93	6.97	*Per cent.* 3.90	....
Beef juice from meat slightly broiled and pressed (neck).....	21.9	9.9	9.4	4.72	5.18	3.56	1.36
Beef tea, chopped beef heated in a bottle without water.........	26.4	7.91	....	5.72	2.19	2.09	....
Beef tea, New England Hospital, with water....................	....	3.23	....	2.55	.68	....	....
Beef tea, with equal weight of water 2 hours at 75° C., then boiled 2 hours..................	....	....	....	....	....	2.15	....
Beef tea with twice its weight of water 2 hours at 70° C., then 2 hours at 85° C..................	....	....	....	....	....	2.62	....
Beef broth, New England kitchen, average of 26 analyses......	....	3.53	....	....	....	4.40	....

It will be seen that the yield of lean beef to water is only about 2 per cent, that is to say, from 3 pounds of juicy steak only about 1 ounce of solid matter is obtained. The broth of the kitchen adds to this 2 per cent from the meat, 2 per cent of gelatin from the bones.

HOW LONG BROTH MAY BE KEPT.

We have taken some pains to determine how long this broth will keep under different conditions, an investigation in the field of household bacteriology. We have settled for our own practice the great importance of: (1) Sterilizing with boiling water every utensil used in the process. (2) Rapid cooling of the broth as soon as made, to below the temperature most favorable to the growth of bacteria. (3) Keeping the broth at as low temperature as possible.

As a result of several experiments during the hot days of July, it was found that broth which spoiled in 12 hours in a cellar where the thermometer stood at 70° F. kept sweet for 7 days if placed while still hot in small jars in an ice chest at 32° F., the cake of fat on the top remaining undisturbed.

Of our many experiments in the cooking of cereals and vegetables we consider only one complete and satisfactory enough to report upon. The principles here demonstrated can, however, be applied to many different dishes.

On account of the high food value of the legumes and the general impression that as ordinarily prepared they are indigestible we determined to make careful experiments on methods of cooking them. First, we undertook to make what would be a standard soup out of dried split pea. We found that in this case also the principle of long, slow cooking was the secret of success. * * * It was found that not until the split pea was cooked 4 or 5 hours, was the result that thick *purée* of rich and mellow flavor that has been so popular in the New England kitchen.

SUGGESTIONS AS TO INVESTIGATIONS NEEDED.

A common remark made by those who have studied the conditions of living of people of moderate incomes and the poor in the different parts of our country is that one of the things most needed for the improvement of the home life of people of these classes is an improvement in their cooking. To replace dear food badly cooked by economical food well cooked is important for purse and health. To make the table more attractive will be an efficient means for making home life more enjoyable and keeping the father and sons from the saloon.

Those who have had experience in cooking schools have come to realize the need of better understanding of the principles which underlie the right preparation of food. The ordinary kitchen stove and the receipt book do not meet the demand, and nothing short of accurate research can meet it.

"The coming fad is domestic science." Indeed that fad is not coming but is already here. The danger is that it may go as a fad and leave no permanently useful impression. But it may be made, instead of a fad, a most salutary educational movement. There is good ground to hope that it will prove so, but only on one condition, namely, that it be based upon thorough scientific knowledge. Such knowledge must be based upon research, just as accurate and just as thorough as that which is given to the study of the other profound problems of chemistry, physics, and hygiene in their application to daily life.

The time is ripe for this inquiry. The leaders in physical, hygienic, and economic science have learned that such subjects are worthy of their attention. The higher institutions of education and research are finding that such inquiry and the teachings that go with them are a part of their truest function. Among the men and women connected with the cooking schools, the experiment stations, the agricultural and mechanical colleges, and other institutions for technical training, the medical schools, and the chemical, physical, and biological laboratories of the universities, there are not a few who are capable of carrying on such inquiries, and with most excellent results.

The general character of the investigations now needed is illustrated by what was said above regarding the digestibility of food and the effects of cooking. More specific suggestions will be best made when the work is ready to be done.

EFFECTS OF COOKING ON DIGESTION.

One of the most important practical topics now demanding investigation is the effect of cooking upon the digestibility and nutritive value of food materials. The general lines of inquiry needed were referred to in connection with the description of experiments above. Some such plans for details as the following are at least worthy of consideration:

Suppose the question to be the effect of the frying of meat upon its digestibility, and that the investigator has at his disposal a subject, a healthy man for instance, who has been trained for the investigation and who is able to attend to some of the mechanical and less agreeable details. A piece of lean beefsteak of proper size could be divided in six parts of like weight and composition and set aside in a refrigerator. Portions representing the whole could be analyzed. The six portions could be eaten, each with given amounts of starch and other carbohydrate materials, and with butter and other substances which would furnish the fats. A ration could be provided which would be fit for sustenance and at the same time would not contain any considerable quantity of material which would interfere with the determination of the actual digestibility of the meat, especially its protein. Three of the portions could be used, during a period of 3 days, nearly raw. The experiment of this period would thus show the digestibility of the meat under the most favorable circumstances; in other words, what might be called its intrinsic digestibility by the subject with whom the experiment was made. The remaining three portions could be fried until well done or overdone, and eaten during a second period of 3 days in the same manner as before. The experiment in this period would show the digestibility of the meat when thus fried. Comparison of the results obtained in the two periods would show the effect of the frying upon the quantity of the meat digested. It must be borne in mind, however, that one series of two comparative experiments would not suffice, but that the investigation should be repeated with the same person and also with different persons. When the experimenter has the method for this kind of work well in hand, it is not extremely laborious, and if conducted for a considerable time will bring results of decided practical value.

This, of course, is only a suggestion for an experiment for testing the effect of one kind of cooking upon one kind of food, but it serves to illustrate how such inquiry may be carried on. Such inquiries are greatly needed, the time is ripe for them, and there are persons who are competent to conduct them. Experience will lead to improvements in methods and indicate the questions which are most desirable for study, and the results will find immediate and widespread application. Here, again, systematic, cooperative work will bring the best results with the least expenditure of time, energy, and money.

CHAPTER VI.

USES OF FOOD IN THE BODY—METABOLISM.[1]

The body of the animal and the food which nourishes it are composed of the same elements. The compounds in the body are more or less similar to those in the food. The processes by which the elements are transformed into the compounds in the plant which serve the animal for food are essentially synthetic. Those by which the compounds of the food are transformed into the compounds of the body are partly synthetic and partly analytic. Those by which the compounds of the food and of the body are utilized in the performance of the bodily functions and are finally excreted are mainly analytic. The generalization of Liebig, that the chemical changes in the plant are synthetic and those in the animal analytic is, in the main, true, but there are important exceptions both in the animal and in the plant.

The later development of science has given us clearer ideas not only of the chemical but also of the physical changes that take place in the living organism.

The physical processes in the plant are mainly those of the transformation of kinetic into potential energy; those in the animal are mainly transformations of potential into kinetic energy. The latent energy of the food is changed to heat and muscular power as the food is used in nutrition.

METABOLISM OF MATTER AND ENERGY IN THE BODY.

To these kinds of change the term metabolism is frequently and appropriately applied. The processes of metabolism in the body are of two definite but closely allied kinds—the metabolism of matter and the metabolism of energy. It is commonly assumed that these two processes conform, the one to the law of the conservation of matter and the other to that of the conservation of energy. Exactly this form of statement is not usual. Indeed, I do not recall having met it in print

[1] The experimental inquiry upon metabolism has been very active during the last few years, and a reasonably full treatment of the subject within the time allowed for the present bulletin is impracticable. It has seemed best, therefore, to present the following brief statements herewith and leave the more detailed treatment for a subsequent publication.

at all, but the principles thus enunciated have been more or less definitely assumed by writers and experimenters during the last 20 years or more. In the following pages the attempt will be made to distinguish between the chemical and physical changes involved in bodily metabolism, and to explain briefly some of the ways by which the changes are measured, and the use that is made of these measurements in the discovery of the ways in which food performs its functions in nutrition.

The bringing of these complex processes into line with the two fundamental laws of the conservation of matter and the conservation of energy helps greatly toward simplifying the whole subject, clearing up details that have been obscure and placing the doctrine of nutrition upon a rational and simple basis. In the light of these laws many of the results of experimenting are more easily interpreted, imperfections in plan and errors in execution of past experimenting are brought out, and the ways in which the unsolved problems before us may best be studied are laid open. Experimenter, teacher, and student are alike helped by this same coordination of principles. At the same time the theory of nutrition thus becomes plainer to the practical man. He can understand it more easily if it is put in terms of "flesh formers" and "fuel values" than if he must consider protein, fats, and carbohydrates in the ways which have become so generally current.

Food has two chief functions—to build tissue and to serve as fuel. In the building of tissue we have to do with the metabolism of matter. In serving as fuel, to yield heat and muscular power, we have to do with the metabolism of energy. The protein compounds are the tissue formers. The fats and the carbohydrates are the chief fuel ingredients; but protein compounds also serve as fuel. In this service as fuel the nutrients replace each other in proportion to their potential energy. The economy of food in nutrition requires sufficient protein for the formation of tissue and sufficient energy for supplying heat and strength.

The fat of the body is its reserve store of fuel. The fuel value of the fats is more than twice as great, weight for weight, as that of the protein or carbohydrates. Fat is body fuel in its most concentrated form. Therein lies the economy of nature in the storage of fat in the body.

The fat of the food is stored as fat in the body. The protein and carbohydrates of the food are transformed into body fat. The fuel value of the food thus becomes a measure of its capacity for fat formation.

Of course the whole doctrine of nutrition is not as simple a matter as these statements would imply, but they do represent its fundamental principles.

In the metabolism of matter and energy we have the foundation of the theory of nutrition, the starting point of experimental inquiry, and the means of simplifying the theories which we have to teach.

METABOLISM OF MATERIAL—DAILY INCOME AND EXPENDITURE OF THE BODY.

To measure the quantities of material metabolized it is necessary to take into account the total income and outgo of matter. The income of matter consists of food, drink, and oxygen of inhaled air. Part of this material is transformed into blood, muscle, fat, bone, and other tissues. The rest, together with the materials worn out with use, undergo various chemical transformations and are finally given off from the body in the form of gases through the lungs and skin, of liquids and solids excreted by the kidneys and skin, and undigested residue of food and metabolic products excreted by the intestines. The substances excreted in these ways constitute the outgo of material. In accurate experiments upon the income and outgo these materials are measured by their chemical elements. It is assumed that all of the gain or loss of nitrogen represents the food and body protein. It is customary to take the quantity of nitrogen multiplied by the factor 6.25 as the measure of this protein. It is also assumed that all of the nitrogen of the income is that of the protein of the food; that all of the nitrogen of the outgo is contained in the nitrogen excreted by the kidneys and intestines. The difference between the nitrogen taken in and that excreted is the measure of the gain or loss of body nitrogen. If there be less nitrogen excreted than was contained in the food, this nitrogen has been stored in the body. Assuming that all has been stored as protein, we have simply to multiply the quantity by the factor 6.25 and we have the quantity of protein stored in the body. In like manner, if there be loss of nitrogen, this quantity multiplied by the same factor gives the loss of protein from the body during the same experiment.

The protein contains besides nitrogen certain quantities of carbon, oxygen, and hydrogen. The assumed average composition of protein will give the quantities of these elements stored in the body as protein or lost in that form. The nitrogen of income and outgo is thus used as the measure of the gain or loss of body protein and also of the gain or loss of carbon, oxygen, and hydrogen in that body protein.

For determining the gain or loss of protein, we next consider the gain or loss of fat. This is comparatively simple, if we assume that there is no gain or loss of carbohydrates. In this case we have need only to use the figures for gain or loss of carbon. Part of the carbon goes with the nitrogen of the protein. If, after this has been calculated out, there remains a loss of carbon, the inference is that the body has lost fat, and as the composition of the fat in the body is practically constant, it is very easy to calculate from the loss of carbon the quantity of fat which the body has lost. In like manner, if the income of carbon is more than the outgo the inference is that carbon has been stored as fat and the quantity of fat thus stored is easily calculated.

In what was just said it was assumed that the quantity of carbohydrates in the body remains constant. To determine whether this is actually so or not, the exact gain or loss of hydrogen and oxygen, as well as of carbon, should be known. Unfortunately experimental inquiry has not yet attained to the degree of perfection which makes it possible to determine the hydrogen and oxygen with the desired accuracy. The determinations of income and outgo of nitrogen are made very easily, since practically all of the nitrogen involved, with the exception of the extremely small amount excreted by the lungs and skin, is given off by the kidneys and intestines and can be determined with great accuracy. The quantity excreted by the intestines is a nearly accurate measure of the quantity digested, the only disturbing factor here being found in the metabolic products. The quantity excreted by the kidneys gives an approximately accurate measure of the total quantity metabolized. The determination of gain or loss of protein is, therefore, a comparatively simple matter, and experiments in which these determinations are made are very common.

The larger part of the carbon of the outgo is excreted in the respiratory products. To determine its amount it becomes necessary to measure and analyze the respiratory products. This is accomplished by means of the respiration apparatus. Various forms of respiration apparatus have been devised. The most successful is that of Pettenkofer, which has been somewhat modified by Voit. With this, carbon of inhaled and exhaled air can be determined with quite satisfactory accuracy. The determinations of hydrogen have been less successful. The materials determined have been carbon dioxid and water. In late experiments at Möckern and Göttingen with domestic animals the hydrocarbons have also been determined.

The tabular statement herewith will serve to indicate succinctly the ways in which the chemical and physical changes take place in the body, in so far as they are now understood, and to compare them with the changes that take place in the calorimetric apparatus which is used for the experimental study of the subject. It should be borne in mind, however, that portions of the food are being constantly stored in the body and becoming parts of its tissues and organs, while previously stored material is at the same time being consumed. The materials stored and consumed may, like those of the food, be grouped as protein, fats, carbohydrates, and mineral matters.

Transformation of material and energy in calorimeter and in animal body.

A.—METABOLISM OF MATTER.

Compounds in food.	Chemical products into which the compounds are transformed by—			
	Direct oxidation in calorimeter.	Cleavage and ultimate oxidation in body.		EXCEPTIONS.
Fats, carbohydrates, alcohol, organic acids, etc.	C all to CO_2......... H all to H_2O......... O forms CO_2 and H_2O	C appears as CO_2......... H appears as H_2O........ O appears as CO_2 and H_2O	Of respiratory products.	1. In respiratory products: Minute quantities of C and H, and perhaps N, P, or S, are given off in hydrocarbons and other volatile compounds. 2. In the excretion of kidneys (and skin): Considerable amounts of C, H, O, N, S, and P appear in urea, uric acid, creatinin, alcohol, etc. 3. In the excretion of the intestine (which includes considerable amounts of undigested or partly digested residues of the food, and with them more or less of metabolic products, mainly from the digestive juices) C, H, O, N, S, and P appear as: *a*, Undecomposed proteids, fats, carbohydrates, etc. *b*, Cleavage products, e. g., salts of organic acids, taurin, skatol, CO_2, CH_4, H, and N?.
Proteids, amides, etc.	C all to CO_2......... H all to H_2O.........	C appears as CO_2 .... H appears as H_2O....	Of respiratory products.	
	S all to $[H_2]$ SO_4..... P all to $[H_3]$ PO_4..... N becomes free N?	S appears as SO_4[1] ...... P appears as PO_4....... N appears as urea, etc.	Excreted by kidneys and skin.	
	O forms compounds as above.	O enters into compounds as above.		
Creatin and allied compounds.	C, H, N, O As above	Not metabolized. (?)		

[1] In combination with bases, metallic or organic.

B.—METABOLISM OF ENERGY.

The potential energy of the compounds of the food is transformed by—		
Direct oxidation in calorimeter—	Cleavage and ultimate oxidation in body—	
Completely into heat.	Mostly into—	But part remains untransformed in the partially oxidized or unoxidized compounds given off from:
	1. Heat..................	1. The lungs and skin in respiration, viz (very minute quantities of): Hydrocarbons and other volatile compounds.
	2. Muscular power	2. The kidneys (and to very slight extent from the skin), viz: Urea, uric acid, creatinin, or other compounds.
	3. Intellectual and nervous energy.(?)	3. The intestine, viz: Undigested residues, and cleavage products of food, and metabolic products, from digestive juices, etc.

In ordinary experimenting the metabolism of mineral matters is not taken especially into consideration, as the quantities of mineral matter and energy metabolized are extremely small.

The total quantity of carbohydrates in the body tissues is also small, and is, like that of the mineral matters, assumed to be very nearly constant. Hence the metabolism of the carbohydrates belonging to the body is ignored in ordinary experiments upon the income and outgo of the body. For accurate results, however, it is essential to determine whether the body loses or gains carbohydrates.

The protein and fats of the body are constantly increasing or decreasing; in other words, the metabolism of body protein and body fat has to be accurately measured and taken into the account in the balancing of the incoming and outgoing of material. The total transformation of energy in the body includes that of body material as well as that of food. It is assumed that the potential energy of body protein and body fat is the same as that of protein and fat of the food. No way has yet been devised for measuring the exact quantities of body protein and body fat metabolized in a given time. What experiments on income and outgo of material do show is the total gain or loss of material in the body. That is to say, if the body during an experiment of 24 hours contains a gram more of fat than at the beginning, the inference is that whatever may have been the quantity of body fat metabolized the quantity of fat stored was equal to this and one gram more. If, on the other hand, the body has lost fat, this loss measures the excess of metabolized fat over that of fat stored from the food, but in neither case do we know the amount of body fat metabolized. What we do learn is the amount of this difference.

The determination of the income and outgo of protein may be illustrated by an actual experiment.[1] The question was this: From a given quantity of the protein of muscular tissue how much will be digested by a healthy man, and will the quantity digested suffice to maintain the supply of protein in his body? In other words, will the man gain or lose protein, or will he simply hold his own on this diet? The subject was a medical student. For protein he ate, in the case here cited, very lean beefsteak. This contained, along with protein, a very small quantity of fat in the form of minute particles which could not be removed with shears and forceps. The diet consisted of the beefsteak cooked with butter, seasoned with pepper, salt, and Worcestershire sauce, and taken with water, beer, and wine. The experiment lasted 3 days. The total quantity of nitrogen of the food was somewhat over 39 grams, of which 38.5 grams were digested, while the rest was excreted in the undigested residue with a small amount of metabolic nitrogen which is not taken into account here. Thirty-seven and two-tenths grams were excreted daily by the kidneys. Assuming that the digested nitrogen of the food represents the income of protein, the gross income of nitrogen, then, is represented by the digested nitrogen of the food, the gross outgo of digested nitrogen by that excreted through the kidneys. Each of these quantities of nitrogen corresponds to a definite quantity of protein which we assume to be equivalent to $6\frac{1}{4}$ times the weight of the nitrogen. It is commonly assumed that for every gram of protein there will be about $4\frac{1}{3}$ grams of muscle, tendon,

[1] W. O. Atwater. Ueber die Verdaulichkeit des Fischfleisches. Ztschr. Biol., 1888, p. 16.

etc., in the meat, and that there will be the same ratio between the weight of muscular tissue consumed in the body and the protein, the nitrogen of which is excreted by the kidneys. That is to say, for every gram of digested nitrogen of the income and outgo there will be $6\frac{1}{4}$ grams of protein and ($6\frac{1}{4}\times4\frac{1}{3}$) about 27 grams of muscle, exclusive of fat.

The gain, then, may be put thus. On a diet of 2 pounds and 10 ounces of lean meat and an ounce of butter per day, was the store of protein in this man's body increased or decreased? In other words, so far as muscular tissue is concerned, did he gain or lose or hold his own? Here are the figures:

Income and outgo of digested nitrogen in experiment with a man on diet of lean meat.

	Grams.
Total nitrogen (per day)	38.5
Nitrogen, kidneys (per day)	37.2
Balance stored in the body (per day)	1.3

That is to say, this young, vigorous man, a student, at his ordinary occupations, studying in his room, listening to lectures at the university, working several hours each day in the laboratory, walking a little for exercise, and living on a diet of protein with a very little fat, gained nitrogen at the rate of 1.3 grams per day. These 1.3 grams of nitrogen represented about 8.2 grams of protein or 35 grams ($1\frac{1}{4}$ ounces) of muscle gained per day during the three days of the experiment. In other words, so far as the lean flesh in his body was concerned he just a little more than held his own.

The difference, that is to say, the gain or loss of fat or of portein, is determined by comparison of income and outgo of material, and thus gives the resultant metabolism of body material. This resultant metabolism of body material expressed in the form of gain or loss of protein or fat must be used in measuring the metabolism of energy. As the factors for this measurement, we may have the total potential energy of the food, the energy of material gained by the body or lost from it, and the energy given off from the body in the form of heat radiated or external mechanical work performed. The income of energy will include the potential energy of the food, and the outgo of energy that of the heat radiated and exterior mechanical work done. If there is a gain by the body in the form of protein or fat; that is to say, if it stores from the food more protein or fat than it consumes of its own substance, the outgo of energy will be diminished by the potential energy of the protein or fat thus stored. If, on the other hand, the body loses protein or fat; that is to say, if it consumes more of its own material than it stores from the food, the potential energy of the material thus lost will go to increase the energy of the outgo. It is clear, therefore, that if we are to study the energy of income and outgo accurately, we must not only determine the energy of the food and that of the heat radiated and exterior mechanical work done, but also the energy of the body material gained or lost. The amount of material gained or lost is determined by comparison of the income and outgo of nitrogen, carbon, and other elements.

THE RESPIRATION APPARATUS.

The experiment above described sufficed to show the nitrogen and protein balance. It showed that the diet of meat and butter with protein and fat sufficed not only to keep up the store of protein in the body for the three days, but also to slightly increase it; but the experiment does not show whether these nutrients supply the body with heat and muscular energy, or whether some of the fat of the body was consumed to make up for the deficiency in the diet.

FIG. 5.—Pettenkofer's respiration apparatus.

The only way to answer this question is to measure the income of other elements and especially of carbon. For this purpose the respiration apparatus is used.

To attempt a description of all of the forms of respiration apparatus which have come into use during the past 40 years and the results of the experiments made with them would far exceed the limits of the present article. The most interesting form and the one which has thus far been most useful is that of Pettenkofer. The following description of this apparatus and of some of the experiments performed with it though originally intended for a more popular exposition, will perhaps suffice for the present purpose:[1]

[1] The description and illustrations are taken from an article by the writer in the Century Magazine for June and July, 1887.

The respiration apparatus is a device for measuring the respiratory products. Many forms have been devised, from one in which the products of respiration of a piece of muscle taken from an animal just killed can be measured, the respiratory process being maintained by artificial circulation of blood through the muscle, to one in which an ox may be kept for days or weeks, and the composition of the inhaled and exhaled air likewise determined.

A very interesting form is that used by the French experimenters, Regnault and Reiset a number of years ago. This was a small chamber of glass, inside of which the animal was placed, some rather complicated appliances being used to continually renew the supply of oxygen and remove the carbonic acid and other products of respiration. But from insufficient ventilation and other minor difficulties this form of apparatus has not quite sufficed for satisfactory experiments, especially with the larger animals and with man.

FIG. 6.—Pettenkofer's respiration apparatus. Pumping machinery.

By far the most satisfactory apparatus is that invented by Professor Pettenkofer, of Munich. This is one of the most interesting devices of modern experimental science. The first one was built through the munificence of the King of Bavaria.

The peculiar features of this apparatus are that the subject of experiment, be it a dog, an ox, or a man, is in a comfortable, well ventilated room, and that the air, which passes through it in a continuous current, is measured and is analyzed both before it goes in and after it comes out. We can thus tell just what the animal has added to it—in other words, what material has been given off as gas or vapor from the body. The arrangements do not provide for estimating all the respiratory products with absolute exactness, but they suffice for reasonably accurate results. The form used for experiments with man consists of a chamber—a "salon," it is called: as a matter of fact it is an iron box—through which a current of air is drawn by a large pump, the latter being worked by an engine.

The salon of the large apparatus at Munich is made of plates of iron, similar to boiler iron, and is in the form of a cube about eight feet each way. It has glass windows, and a door large enough to admit a man. The large engraving (fig. 5, p. 106) shows the apparatus as it is now arranged. On the left is the chamber in which the man under experiment stays; near are a table holding apparatus for analyzing the air before and after it passes through the chamber, and a large meter for measuring the quantity of air which passes through. In an adjoining room is the machinery by which the current of air is pumped through the apparatus. The smaller sketch

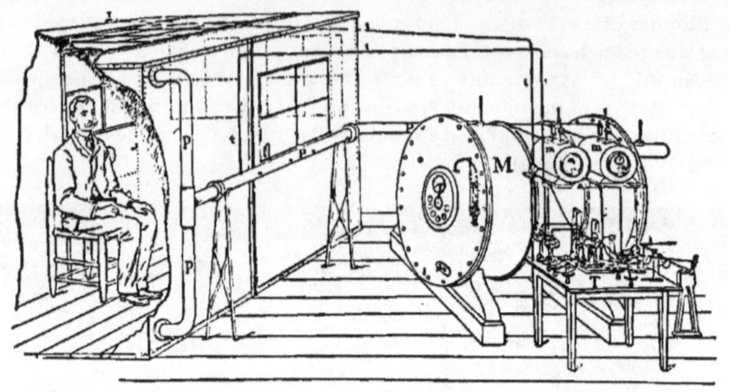

FIG. 7.—Pettenkofer's respiration apparatus. Explanatory sketch.

explains the working in more detail. The air enters the chamber at its left side and passes out on the right through the large pipe P P into the large meter M, in which it is measured. A small tube *t t* takes from the pipe P P a portion of the air, which has been passed through the chamber and contains the products of respiration, into two small meters *m m*, where it is measured, and through the apparatus on the the table T, where it is analyzed. A similar small tube *t′ t′* brings air for analysis

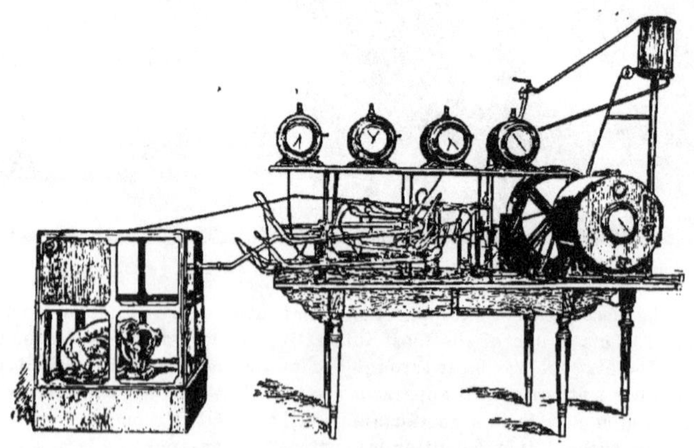

FIG. 8.—Voit's respiration apparatus.

from the outside of the apparatus, taking it from the left of the chamber where it enters the latter and carrying it into two other small meters (not shown in this sketch), where it is measured, and through apparatus, also not shown here, by which it is analyzed. In the engraving (p. 106) the four small meters and apparatus for analyzing the air are shown on the table between the chamber and the large meter. Comparisons of the quantity and composition of the air which has passed through the chamber with the outside air show what the man has imparted to the air in breathing, and thus tell the amounts of the products of respiration. The food and drink and the solid and liquid products of its consumption in the body are at the same time measured, weighed, and analyzed, and thus all of the items of income and outgo of the body are determined.

Figure 9[1] represents a smaller form of respiration apparatus devised by Voit. It is identical in principle with the larger apparatus. It is intended for experiments with dogs, geese, and other small animals. Its object is to provide for analysis of the air before and after it has been breathed by the animal, and thus show what products of respiration the animal has imparted to it. The box in which the animal is kept is made of glass. Through this box a constant current of air is drawn and measured by the large meter on the table. A small portion of this, however, is drawn through two of the small meters by which it is measured, and through apparatus on the table by which it is analyzed. Air taken from outside the box is at the same time drawn through the other two small meters and apparatus on the table, and thus measured and analyzed in like manner.

The first man to enter the respiration apparatus for experiments upon himself, I believe, was Professor Ranke, of Munich, who has described his experiences in his book on The Nutrition of Man (Die Ernährung des Menschen), as well as in special memoirs. He tells us that in trials in which he took no food the fasting was somewhat disagreeable, but far less painful than many would think. "I found myself at the end of the first 24 hours entirely well; at the end of the second 24 hours without food or drink, during which sleep had been disturbed, the head was somewhat heavy and there was an oppressiveness in the stomach and considerable weakness; but the sensation of hunger, * * * which was strongest about 30 hours after the last food was taken, * * * did not appear any more."

In the greater number of Professor Ranke's experiments he took a reasonable amount of food. The diet was simple, and consisted of such materials as lean meat, bread, white of egg, starch, sugar, butter, etc., and was found to serve the purpose very well. After some experience a ration was arranged which corresponded very well in composition with that used by ordinary working people, and was at the same time not at all unacceptable. When a number of experiments with Professor Ranke had been completed, several series were made with other persons. One of these latter series I will briefly describe.

The subject was a strong, healthy mechanic, a watchmaker, 28 years old, and weighing about 156 pounds. Three experiments were made, each occupying 24 hours. In the first, the man took nothing but a little meat extract, salt, and water, and did no work. In the second, he had a liberal allowance of palatable food, but still remained at rest. In the third, he had the same diet as in the second, but worked hard at turning a lathe for 9 hours, so that he was thoroughly tired at night. During the daytime of the first two experiments, I should say, he read, cleaned a watch, and otherwise occupied himself to while away the time, making, however, very little muscular effort.

The three experiments, then, show the effects of fasting and rest, food and rest, and food and muscular exercise upon the income and outgo of this man's body. We will note only very briefly some of the details of the experiments, the full accounts of which fill many pages.

The diet of the first experiment consisted of: Meat extract, 12.5 grams (a little less than one-half ounce); salt, 15.1 grams (a little over one-half ounce); water, 1027.2 grams (about a quart).

The day's ration of the second trial included a third of a pound of lean meat, a pound of bread, a little over a pint of milk, and about a quart of beer, and other materials as follows:

---

[1] Century Magazine, XXIV, 1887, p. 402.

Day's food in second experiment.

	Grams.
Meat, lean beef	140
Egg albumen (white of egg)	42
Bread	450
Milk	500
Beer	1,025
Lard	70
Butter	30
Starch	70
Sugar	17
Salt	4
Water	286

The diet of the third experiment was essentially the same as that of the second, except that the man drank a little more water.

The income included, besides the food and drink, the oxygen consumed from the inhaled air. The estimated quantities were:

Oxygen used in 24 hours.

	Grams.
First experiment, fasting and at rest	779
Second experiment, liberal ration and at rest	709
Third experiment, liberal ration and at work	1,006

The final balance sheets of the experiments, which show the details of income and outgo in terms of the chemical elements, carbon, nitrogen, etc., are too extensive to be reported here. That for each experiment would nearly fill one of these pages, but as some readers may be curious to see what they are, I give the principal data in abbreviated form:

Daily income and expenditure of chemical elements.

	Carbon.	Hydrogen.	Nitrogen.	Oxygen.
Experiment with no food (except meat extract) and no work:	*Grams.*	*Grams.*	*Grams.*	*Grams.*
Income	2.4	115.1	1.2	1698.4
Outgo	209.5	221.6	12.5	2301.4
Loss	207.1	106.5	11.3	603
Experiment with liberal ration of meat, milk, bread, etc., and no work:				
Income	315.5	270.9	19.5	2712.9
Outgo	275.7	248.2	19.5	2630.2
Gain	39.8	22.7	0.0	82.7
Experiment with liberal ration, as in preceding experiment, and hard work:				
Income	309.2	297.7	19.5	3232.5
Outgo	336.3	304.9	19.5	3246.5
Loss	27.1	7.2	0.0	14

But we wish to know what quantities of flesh and fat the man gained or lost under these different conditions of food and fasting, labor and rest. The figures just cited are for the chemical elements of which the protein and fats are composed. Knowing the proportions of the elements in each compound, it is easy, from the figures for the elements, to estimate the quantities of the compounds. Omitting details of the calculations the results are given in the balance sheet of compounds herewith. Regarding the carbohydrates, however, I should explain that since the body has extremely little of its own, and those of the food are consumed, they are left out of account in the experiment without food, and the amounts received and consumed in the experiments with food are taken as balancing one another.

Income and expenditure of chemical compounds by body of man.

	Fasting (no work).			Liberal ration (no work).			Liberal ration (hard work).		
	Protein.	Fats.	Carbohydrates.	Protein.	Fats.	Carbohydrates.	Protein.	Fats.	Carbohydrates.
	Grams.	*Grams.*	*Grams.*	*Grams.*	*Grams.*	*Grams.*	*Grams.*	*Grams.*	*Grams.*
Income	7	0	None.	122	117	332	122	117	352
Outgo	78	216	None.	122	52	332	122	173	352
Gain +, or loss —	—71	—216	None.	0	+65	0	0	—56	0

The protein gained or lost was mainly from the muscles and similar tissues, or what we may call flesh as distinguished from fat. Taking the figures for protein and fats gained and lost as shown in the last line of the balance sheet of income and expenditure of compounds, changing grams to ounces, and assuming that with each ounce of protein would be water, etc., enough to make the equivalent of 4½ ounces of lean flesh, i. e., muscle, tendon, etc., we have this final result of the trials; the quantities, as before, are those gained or lost in one day:

Outcome of the experiments as regards increase or decrease of lean flesh and fat within the body.

	Lean flesh (muscle, etc.).	Fats.
	Ounces.	*Ounces.*
No food, no work, loss	11	7⅔
Liberal diet, no work, gain	None.	2⅓
Liberal diet, hard work, loss	None.	2

That is to say, fasting, and without muscular labor, the man lived upon the tissues of his body, and consumed daily a trifle less than three-quarters of a pound of muscle, and with this nearly half a pound of the fat previously stored in his body. With plenty of food, and still resting, he neither gained nor lost lean flesh, but gained 2⅓ ounces of fat in a day. And when he set himself to hard muscular work, with the same amount of food, he likewise held his own so far as lean flesh was concerned, but lost 2 ounces of fat. The body used for its support protein and fats, in each case, and carbohydrates when it had them. When the nutrients were not supplied in food, it consumed a little protein and a good deal more fat from its own store. With a ration which sufficed to exactly maintain its protein without gain or loss, the body gained fat when it had only a little more than its own muscular work to perform (that included in breathing, keeping the blood in circulation, etc.), and lost fat again when this work was increased by manual labor.

If we had only these experiments to judge from, we might infer that muscular energy comes from consumption of fat, and that the special work of the protein of the food is to repair the wastes and make up for the wear and tear of the protein of the body; and this would be true as far as it goes. But, of course, many other experiments and of many different kinds are needed to settle these questions. The majority of the most useful ones, thus far, have been made with other animals than man.

In studying the laws of animal nutrition the most convenient organism, for many purposes, is that of the dog. The dog thrives upon both animal and vegetable foods, utilizes large quantities of food to advantage or endures long fasting with patience, and makes ready responses by changes of bodily condition to changes in the food. In reading the accounts of the famous feeding trials conducted by Bischoff and Voit, one is surprised to see what control they obtained of the organisms of the dogs experimented with. By altering the kinds and quantities of food constituents, Voit was able either to reduce both the flesh (protein) and the fat of the animal's body or to increase both flesh and fat, or to reduce the one or to increase the other. Indeed, the manipulations effected in this way seemed almost equivalent to getting into the tissues and directly removing or adding flesh or fat at will. The principles thus learned from experiments with the dog and other animals apply in the main, though not in all the details, to the nutrition of man.

The effect of one-sided diet is very well illustrated in some experiments by Professor Ranke. They were made in the respiration apparatus at Munich, and belonged to the series of which I have already spoken. After he had studied the changes that went on in his body when fasting, he proposed to himself these questions:

What will be the effect of a diet of protein with very little fat and no carbohydrates on the one hand, and of a diet of fats and carbohydrates without protein on the other? In other words, how will the composition of the body be affected by food rich in protein and containing little else, and how will the store of fat and protein be altered by leaving the protein out of the food and living on the other nutrients?

For the diet of protein, he took lean meat, with butter and a little salt, essentially the same diet as was used by the student in the experiment described above. He had found himself able to eat 2,000 grams of the lean meat in the course of the day, but in this experiment, which lasted 24 hours, he ate only 1,833 grams (about 4 pounds) of meat and with it 70 grams of fat, 30 grams of salt, and 3,371 grams (nearly 3 quarts) of water. Without going into the details, suffice it to say, that, according to Professor Ranke's calculations, his body lost 15.1 grams of fat and at the same time gained 113 grams of protein during the day of the experiment. In the other experiment, which likewise continued for 24 hours, the food consisted of 150 grams of fat, 300 grams of starch, and 100 grams of sugar, an even less appetizing mixture perhaps than the lean meat and butter for an exclusive diet, but yet one which, if put together with proper culinary skill, makes a cake that can be swallowed. This time he lost 51 grams of protein and gained 91.5 grams of fat.

The results of these two experiments may be recapitulated thus: On the diet consisting chiefly of protein (lean meat, etc.) the body gained protein (muscle, etc.) and lost fat. On the diet consisting chiefly of fats and carbohydrates (starch and sugar) the body lost protein and gained fat. This is just what we might expect. But it is interesting to have the facts and figures to show exactly what did take place, and other experiments make it safe to say that if either the quantities of food or the condition of Professor Ranke's body had been different, the results would have been different also. Thus in the first experiment if he had eaten less meat he would have stored less protein; indeed, with a small enough ration he would have lost both protein and fat, and it seems probable that if he had not been a rather fat person he would not have lost fat so readily on the protein diet.

Experiments confirm and to some extent explain the fact so well attested by general experience, that a mixed diet is best for ordinary people in health. Professor Ranke found that when he did no muscular labor, his body neither gained nor lost; that, in other words, he just about "held his own" with food, containing per day: Protein, 100 grams (3.5 ounces); fats, 100 grams; carbohydrates, 240 grams (8.5 ounces).

CHAPTER VII.

METABOLISM OF ENERGY—INCOME AND OUTGO OF BODY.

In considering the metabolism of energy we have to deal with the principles of thermo-chemistry as applied to animal and vegetable organisms. For the present purpose we may confine our attention to the changes which go on in the animal body. The principles to be applied may be expressed in various ways. They are enunciated by Berthelot in the following terms:[1]

I. *Principle of molecular [atomic] work.*—"The quantity of heat evolved is the measure of the sum of the chemical and physical work accomplished in any reaction."

II. *Principle of conservation of energy.*—"When a system of bodies—simple or compound—starting from a given condition undergoes either physical or chemical changes, which bring it into a new condition without producing any mechanical effect on external bodies, the amount of heat evolved or absorbed, as the total result of these changes, depends solely on the initial and final states of the system, and is the same, whatever may be the nature or order of the intermediate states."

III. *Principle of maximum work.*—"In any chemical reaction between a system of bodies not acted on by external forces the tendency is toward that condition and those products which will result in the greatest evolution of heat."

It is with the second and third of these principles, and especially the second, that we have to do. In discussing this subject Professor Cooke says:

> We readily accept Berthelot's second fundamental principle of thermo-chemistry when enunciated as above, because it so obviously falls under the general law of conservation of energy; but it is obvious that this principle could not have been assumed prior to its experimental verification any more than could the principle of the conservation of mass prior to the experiments of Lavoisier, and as Lavoisier worked out this last great principle with the balance, so Berthelot and Thompson have demonstrated with the calorimeter the corresponding fundamental principle of thermo-chemistry, which must be regarded as a generalization from the results of their work. Moreover, although in cases of simple direct combination the principle under discussion is almost self-evident and has been long admitted, yet before the investigations of Berthelot and Thompson no chemist conceived of its application in the very complex and indirect reactions by which the greater part of the thermo-chemical data have been obtained.

In like manner it may be said that the application of the second and third of these principles, and especially the second, in the living organism has long been believed, but has lacked the absolute demonstration. During the past ten years, however, an approach to such demon-

[1] Essai de mécanique chimique, Tome I, Introduction, pp. xxviii, xxix. See also J. P. Cooke in Amer. Jour. of Sci., 3d. ser., XIX, pp. 261-267; and Pattison-Muir, Thermal Chemistry, pp. 297 and 245, note.

stration has been made by several experimenters, notably by Rubner in experiments with dogs, and although these have not the completeness necessary to place the principle beyond all peradventure, and especially to show the details of its application, they suffice to confirm the belief in the general application of the principle.

The experimental study of this question is carried on in two lines.

Potential energy—Heats of combustion determined by the bomb calorimeter.—The first of these kinds of inquiry has to do with the determining of the potential energy of the materials concerned in the metabolic processes. These materials are (1) the nutrients, i. e., protein, fats, carbohydrates, etc., of the food; (2) the substances, mainly protein and fats, which are either stored in the body or taken from the store contained in the body and consumed; and (3) the excreted compounds which are not completely oxidized and which in consequence still contain potential energy; these are chiefly the urea and other organic compounds excreted by the kidneys and the undigested residue of the food. The heats of combustions of the compounds, as determined by oxidation in the bomb calorimeter, are taken as the measure of their fuel value.

Income and outgo of energy in the body—Respiration calorimeter.—The second branch of the inquiry concerns itself with the balance of energy in the body. It involves the determination of the total income and outgo expressed in terms of both matter and energy. The income and outgo of matter are determined by the aid of a respiration apparatus, which shows the quantities of chemical elements and compounds taken in and given off by the body, and inferentially the quantities of material which are either stored in the body during the experiment or consumed from its previous store. The income of energy is represented by the potential energy of the materials which constitute the income of matter. Their potential energy is learned from the heats of combustion developed when they are burned with oxygen in the calorimeter. The outgo of energy is made up essentially of three factors. One is the potential energy of the excreta, which is determined by burning with oxygen in the calorimeter; another and the principal one is the heat radiated from the body. To measure this, different appliances have been used. The most successful results thus far published have been obtained by Rubner with a respiration apparatus which has special arrangements for determining the quantity of heat radiated from the body. Perhaps the most convenient designation for such forms of apparatus is that here used—respiration calorimeter. The third factor is the mechanical work done by the body, i. e., that which manifests itself as external work, and not including the internal work such as is involved in respiration and the circulation of the blood. The heat equivalent of this external mechanical work, added to the heat radiated from the body and the heat of combustion of the excreta, would make up the total outgo of energy. No successful attempt to measure all these factors in the same experiment has yet been published.

In the study of the nutrition of man in relation to his health and useful work the sources, nature, and economy of intellectual energy constitute a most important factor, but one upon which chemical physiology has as yet thrown very little light.

POTENTIAL ENERGY OF FOOD—HEATS OF COMBUSTION—FUEL VALUES.

The food performs essentially two functions in the body—the building and repair of tissue and the yielding of energy. In being consumed the nutrients yield energy in the form of either heat or muscular power. Part of this potential energy becomes kinetic in the cleavage of complex compounds to simpler ones; part is liberated in the processes of oxidation. Neither the chemical nor the physical changes which take place are now fully understood. Of this much, however, we are certain: The processes are complex, and although the ultimate chemical products may be the same as those of direct oxidation, the processes by which they are formed in the body are much more complex than those which take place when they are burned either in the furnace or the calorimeter. But it is believed that, in accordance with the principle of the conservation of energy, the quantity of potential energy which is transformed into kinetic energy will be the same in the one case as in the other, provided the final products are the same. Furthermore, in accordance with the principle of maximum work the tendency is toward those changes which result in the greatest evolution of heat or other form of kinetic energy. We may therefore take the heats of combustion of the nutrients of the food as equivalent to their potential energy; we may also take this potential energy as the measure of their fuel values, i.e., their value for the production of heat and muscular work when they are consumed in the body. The same principle applies to the materials, mainly protein and fats, which the body takes from the food and makes a part of its tissue before they are consumed. It applies also to the incompletely oxidized excretory products like urea and to the undigested residue of the food which is excreted by the intestine, in so far as their potential energy is concerned. That is to say, if we subtract the potential energy of these products from that of the total material from which they are formed the difference will be the amount of energy which has been liberated in their consumption in the body. If, however, a portion of the food has been stored as part of the tissue of the body, the potential energy of the compounds thus stored must also be subtracted from that of the total food in order to learn the quantity of energy actually liberated. Speaking in general terms, then, the quantity of heat which is set free when a given food material is burned with oxygen in the calorimeter may be taken as the measure of the fuel value of that food material.

Such is the theory. It needs further experiment for complete demonstration, but the balance of probability is now very strongly in its favor, and growing more and more so as accurate research accumulates. It becomes very important, therefore, to determine accurately the heats of combustion of the ingredients of food and their metabolic products,

CALORIMETERS AND CALORIMETRY—HISTORICAL DEVELOPMENT.

The history of calorimetry as applied to determining the heats of combustion of organic substances is too comprehensive a subject for treatment here, but a brief review of the part of it which bears most directly upon the fuel values of the compounds concerned in metabolism will not be out of place.

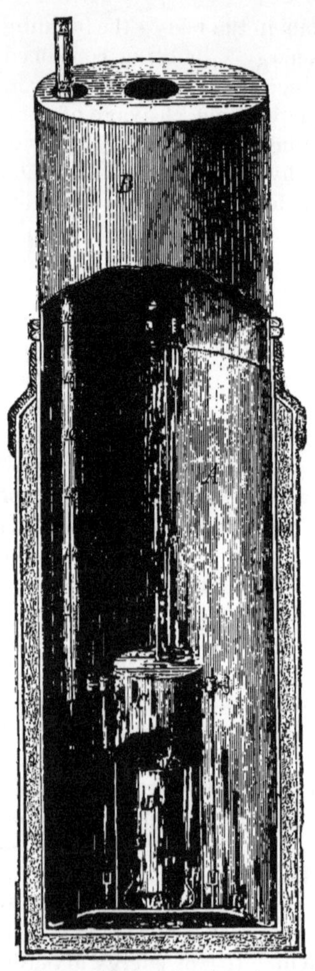

FIG. 9.—Thompson's calorimeter.

THOMPSON'S CALORIMETER—INVESTIGATIONS OF FRANKLAND.

In determining the heats of combustion of such compounds two distinct forms of apparatus have been employed. In both the substance is burned with oxygen and the heat is measured by the rise of temperature of a certain weight of water to which the heat is communicated. The older one is that of Lewis Thompson.[1] This was used for determining the heats of combustion of organic compounds as early as 1866.[2] The apparatus consisted eventually of a stout copper cylinder closed at the bottom and standing upright inside a so-called "diving bell," also of copper. The cylinder and diving bell were immersed in a suitable vessel containing 2 liters of water. In performing the experiment a given weight, about 2 grams, of the substance, whose thermal value was to be determined, was intimately mixed with 19.5 grams of potassium chlorate, to which about 2.5 grams of manganese dioxid had been added. This mixture was placed in the cylinder; a small piece of cotton, previously steeped in a solution of potassium chlorate and dried, was inserted as a fuse. The temperature of the water was then carefully ascertained by a delicate thermometer, and, after the end of the cotton thread had been ignited, the tube, with its contents, was placed in the

[1] Described briefly by Frankland, Proc. Roy. Inst. of Great Britain, June 8, 1866, and Phil. Mag. [4], 32, 182.

[2] Loc. cit. See also résumé of subject by Stohmann, Jour. prak. Chem. (N. F., 19), 1879, 115.

bottom of the copper diving bell and lowered to the bottom of the water. When the mixture burned the gaseous products of combustion issued from numerous small openings at the bottom of the bell and rose to the surface of the water—a height of about 10 inches. At the end of the deflagration the water was admitted to the diving bell and replaced the remainder of the gaseous products, which were allowed to escape through a small tube. The water was well mixed by moving the bell up and down repeatedly, and its temperature was again carefully observed. The rise in temperature of the water, compared with its specific heat, gave the quantity of heat liberated in the combustion. Corrections were applied for (1) the heat absorbed by the apparatus, and (2) the heat evolved by the decomposition of the chlorate of potassium, the former being added to and the latter subtracted from the heat given off in the combustion, as measured by the rise in the temperature of the water. Both corrections were determined experimentally once for all, and the values thus obtained were applied in each determination of the heats of combustion of the substance studied. A series of determinations were made by Frankland, who states the results as follows. The heat units are small calories:

Actual energy developed by 1 gram of each substance when burnt in oxygen.

[Heat units.]

Name of substance (dried at 100° C.).	Experiment.				
	First.	Second.	Third.	Fourth.	Mean.
Beef muscle purified by repeated washing with ether	5,174	5,062	5,195	5,088	5,103
Purified albumin	5,009	4,987	....	....	4,998
Beef fat	9,069	....	....	....	9,069
Hippuric acid	5,330	5,437	....	....	5,383
Uric acid	2,645	2,585	....	....	2,615
Urea	2,121	2,302	2,207	2,197	2,206

The special purpose of these investigations was to get information regarding the quantity of energy which could be developed in the body from the consumption of protein. With reference to this question, Frankland makes the following statements and calculations:

It is evident that the above determination of the actual energy developed by the combustion of muscle in oxygen represents more than the amount of actual energy produced by its oxidation within the body, because when muscle burns in oxygen its carbon is converted into carbonic anhydrid, and its hydrogen into water, the nitrogen being to a great extent evolved in the elementary state; whereas when muscle is most completely consumed in the body the products are carbonic anhydrid, water, and urea; the whole of the nitrogen passes out of the body as urea, a substance which still retains a considerable amount of potential energy. Dry muscle and pure albumin yield, under these circumstances, almost exactly one-third of their weight of urea; and this fact, together with the above determination of the actual energy developed in the combustion of urea, enables us to deduce with certainty the amount of actual energy developed by muscle and albumin, respectively, when consumed in the human body. It is as follows:

Actual energy developed by 1 gram of each substance when consumed in the body.

Name of substances (dried at 100° C.).	Heat units (mean).
Beef muscle purified by ether	4,368
Purified albumin	4,263

Arguing that the source of animal heat and of muscular power is to be found in the transformation of the potential energy of the food into kinetic energy, and that "from this point of view it is interesting to examine the various articles of food in common use, as to their capabilities for the production of muscular power," Professor Frankland "made careful estimates of the calorific value of different materials used as food with the same apparatus and in the same manner as described above." The results are summarized in the following table:

Results of experiments with food dried at 100° C.

[Heat units.]

Name of food.	Experiment.		
	First.	Second.	Third.
Cheshire cheese	6,080	6,149	....
Potatoes	3,752	....	....
Apples	3,776	3,562	....
Mackerel	5,994	6,134	....
Oatmeal (not dried)	4,143	4,018	3,857
Lean beef	5,271	5,260	5,410
White of egg	4,823	4,940	4,927
Carrots	3,776	3,759	....
Pea meal (not dried)	3,866	4,006	....
Flour (not dried)	3,941	3,931	....
Arrowroot (not dried)	3,923	3,962	....
Butter	7,237	7,291	....
Ham, boiled and lean	4,188	4,498	....
Lean veal	4,459	4,595	4,488
Hard-boiled egg	6,455	6,187	....
Yolk of egg	6,460	....	....
Isinglass	4,520	4,520	....
Cabbage	3,809	3,744	....
Whiting	4,520	4,520	....
Ground rice (not dried)	3,802	3,824	....
Cod-liver oil	9,134	9,080	....
Cocoa nibs (not dried)	6,809	6,937	....
Residue of milk	5,066	5,120	....
Bread crumb	3,984	3,984	....
Bread crust (not dried)	4,459	....	....
Lump sugar (not dried)	3,403	3,294	....
Commercial grape sugar (not dried)	3,277	3,277	....
Residue from bottled ale	3,776	3,744	....
Residue from bottled stout	6,348	6,455	....

From the data thus obtained Frankland estimates the quantities of energy actually developed by these materials when oxidized in the body, making allowance for the nitrogenous material which is not completely oxidized. He adds that "it must be borne in mind that it is only on condition of the food being digested and passed into the blood that the results given in these tables are realized, and that the force values experimentally obtained for the different values in these tables, must, therefore, be understood as the maxima assignable to the substances to which they belong." The tables in which the compilations are given are somewhat detailed and need not be repeated here.

STOHMANN'S CALORIMETER AND INVESTIGATIONS.

After the publication of Frankland's investigations in 1866, no further investigations of importance in this direction were announced or undertaken until 1877, when Stohmann began work in Leipsic. His results were first published in 1879. His method was the same in principle as that followed by Frankland. Stohmann, however, gave much attention to the sources of error, and devised a form of calorimeter with which the attempt was made to avoid them.

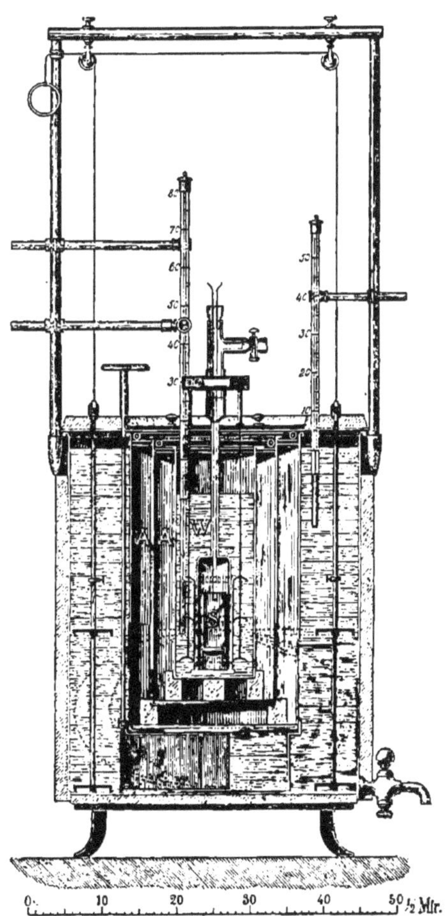

FIG. 10.—Stohmann's calorimeter.

The figures of Frankland above cited show wide discrepancies in the duplicate determinations of the heats of combustion of the same material. Stohmann's first effort was to trace these discrepancies to their source. The principal causes of error were found in the oxidation of the copper of which the cylinder was made and the irregularity of a secondary thermal process incident to the combustion, namely, the solution of the potassium chlorid formed by the decomposition of the chlorate. The former error was obviated by substituting platinum for copper in the combustion cylinder and the latter was controlled by the direct determination of the potassium chlorid in solution at the end of the combustion. Stohmann used at first an apparatus which was obtained in London and was essentially the same as that of Thompson which Frankland had employed. He soon modified it, however, by making the combustion cylinder of platinum instead of copper, as just stated, and by substituting a vessel of brass in the place of one of glass for holding the water.[1]

In the elaboration of the method and finding the sources of error and means of avoiding them, Stohmann, with his two assistants, Dr. Spindler and Dr. von Rechenberg, made "at least a thousand combustions." In the apparatus thus described a layer of nonconducting material (wool) was put around the outside of the brass cylinder which contained the water; no other means was used to prevent the radiation

[1] A description of this apparatus with diagram is given by Stohmann in Jour. prak. Chem., 127 (N. F. 19), 1879, p. 115.

of heat from the water in the cylinder outward, or the passage of heat from without into the water. Later, Stohmann provided a series of three concentric copper cylinders. The space in the interior of the inner cylinder which was not occupied by the apparatus was filled with air; the space between this cylinder and the second was also filled with air; that between the second and the outer one was filled with water. The outer cylinder was wrapped in felt. There were thus 2 layers of air, 1 of water, and 1 of felt between the calorimetric apparatus proper and the external air. Mechanical devices were provided for stirring water in the calorimeter so as to secure uniform distribution through it of the heat from the combustion. The temperature of the water was measured by a thermometer with divisions such as to permit reading to one-thousandth of a degree with the aid of a magnifying glass.[1]

The apparatus thus perfected was used for ten years or more by Stohmann and his pupils, Von Rechenberg and Danilewski in Leipsic, by Rubner in Munich, and by Gibson in the writer's laboratory.

THE BOMB CALORIMETER OF BERTHELOT.

In the combustions by the Thompson-Stohmann method just described the oxygen is obtained from potassium chlorate. Considerable time is required for the determination, but the chief difficulty with this method is that the combustion is not always complete. Berthelot has devised an apparatus and, with the assistance of Vieille and others, has developed a method for the use of oxygen under high pressure. The apparatus consists essentially of a steel bomb lined with platinum, within which the substance is burned. The bomb is immersed in water contained in a metal cylinder; this calorimeter cylinder is placed with its contents inside of concentric cylinders containing air and water. The heat developed is measured by the rise in temperature of the water, due allowance being made for the heat absorbed by the metal of the apparatus and for that introduced in igniting the substance by an electric current and developed by the oxidation of an iron wire through which the current is passed, and in the formation of a small amount of nitric acid. The reactions are simple, the oxidation of the compounds in completing the determination requires a comparatively short time, and the results are very satisfactory. The only drawback is the great cost of the apparatus which is due to the large amount of platinum employed in its construction.[2]

The bomb calorimeter was first used by Berthelot in measuring the heats of combustion of gases by detonation. The gas to be burned was mixed with the exact amount of oxygen required, or a slight excess at ordinary atmospheric pressure within

[1] See description by Stohmann with diagram; Landw. Jahrb., 13, 1884, 513. It is also pictured in the Century Magazine, July, 1887.

[2] For descriptions see Berthelot. Ann. Chim. et Phys. (5), 23, 160; Berthelot and Vieille, Ibid. (6), 6, 546; Berthelot, Traité pratique de calorimétrie chimique, Paris, 1893, p. 128, and, for an especially clear account of the apparatus and method, Stohmann, Jour. prak. Chem., 147 (N. F. 39), 1889, 503. An engraving of the bomb with a brief description may be found in Ztschr. analyt. Chem., 1893, 77. The apparatus employed by Stohmann was obtained from Golaz in Paris, who made it in accordance with Berthelot's directions. It cost, with accessories, including compression pump, etc., not far from $1,200.

the bomb and ignited by the passage of an electric spark. Gases were easily burned in this way, but the combustion of solids was impracticable. Attempts were made to remedy the difficulty with solids by intimately mixing them with potassium chlorate and good results were obtained as in the Thompson-Stohmann method. Berthelot and Vieille[1] found later,[2] however, that when oxygen was introduced under a pressure of from 7 to 25 atmospheres, the combustion was complete with all organic substances even when no potassium chlorate was added.

The bomb originally employed by Berthelot was in the shape of a cylinder, with hemispherical ends and divided in two parts, one of which screwed into the other. It was made of steel with the interior lined with gold by electroplating.

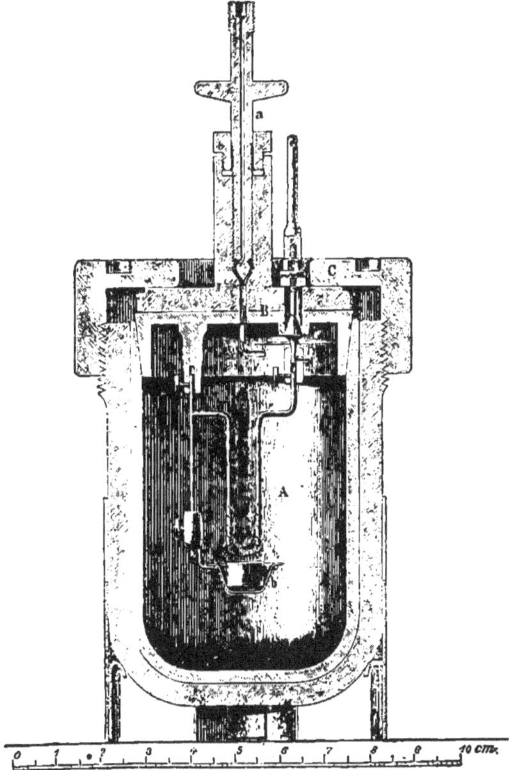

FIG. 11.—Berthelot's bomb calorimeter.

The later and permanent form of Berthelot's bomb is cylindrical with a rounded bottom and flat cover. It is made of steel with a heavy lining of platinum. The size may vary. In the one made by Golaz for Stohmann the diameter is approximately 10 centimeters and the height to the upper surface of the cover about 13 centimeters;

[1] Ann. Chim. et Phys. (6), 6. 546.

[2] In 1885. The principle was discovered and investigated by Frankland in 1864 and 1868. See investigations by him "On the combustion of iron in compressed oxygen." Jour. Chem. Soc., II, 1864; 52. "On the combustion of gases under pressure. Brit. Assoc. Rep. XXXVIII, 1868 (*Sect.*); 37. "On the combustion of hydrogen and carbonic oxid in oxygen under great pressure." Proc. Roy. Soc., XVI, 1868; 419.

the walls are somewhat over a centimeter in thickness. It contains 2,717 grams of steel and 1,233.3 grams of platinum. The internal capacity is about 300 cubic centimeters and it holds, under a pressure of 24 atmospheres, 7 liters, or, in round numbers, 10 grams, of oxygen. The cover fits into the top of the cylinder after the manner of a stopper. It is pressed in and held very tightly by an outer cap or collar which screws on to the outside of the cylinder at the top. The fittings have to be made with the greatest care in order to prevent the escape of gas. The material to be burned is shaped into a small cake by means of a powerful press and is held in a platinum capsule. This capsule is sustained by a platinum wire which is fastened to the under side of the cover. Another platinum wire passes through the cover, from which it is insulated by gutta percha, or other appropriate material. These two wires are so arranged that they are easily connected by a piece of fine iron wire hanging over the substance to be burned in the platinum capsule. An electric current passed through the platinum wire heats the iron wire to a temperature where it burns in the oxygen and, melting, falls upon the substance so as to ignite it. An arrangement at the top of the cover provides for admitting the oxygen. The oxygen is introduced either with the aid of a compression pump, or, more conveniently, from iron or steel cylinders, in which it is held under sufficient pressure. Both Berthelot and Stohmann use such cylinders without the aid of a pump. Experience has shown that it is desirable to have fully three times as much oxygen present as is theoretically necessary for the combustion.

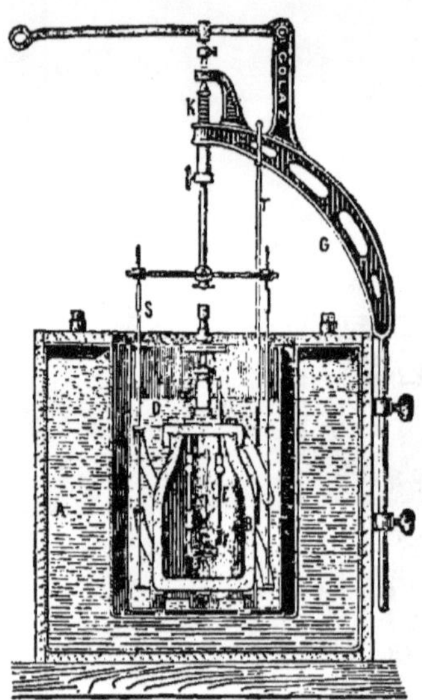

FIG. 12.—Mahler's bomb calorimeter as mounted for use.

The quantity of water in which the bomb is immersed is generally about 2 liters. It is contained in a calorimeter cylinder of brass or other metal. A stirrer, not easily described without a diagram, plays between the bomb and the cylinder in such way as readily to mix the water and insure uniform temperature after the combustion. The outer cylinders which are employed by Berthelot and Stohmann to protect the apparatus from gain or loss by heat outside, as above described, are made of copper. It is found that 2 concentric cylinders so arranged as to hold a layer of water between them, the inner being large enough to leave a considerable air space around the calorimeter cylinder, suffice for accurate work. In Stohmann's laboratory the arrangements to insure accuracy are quite elaborate. The work is done in a basement room surrounded by very thick walls of stone. Special devices are employed to keep the temperature of the room exactly constant. The stirrer is moved by a small motor, which is so regulated as to make the movement the same for all determinations. Berthelot uses a motor for the stirring, but conducts the combustions in the laboratory rooms where other work is done and without special arrangements to insure uniform temperature.

The Berthelot bomb calorimeter serves its purpose admirably. It is comparatively simple, easily handled, and does not get out of order when properly cared for. Practically all kinds of ordinary organic compounds are completely oxidized when the proper excess of oxygen is used at an initial pressure of 25 atmospheres. With an accurate thermometer the rise in temperature of the water is measured with great accuracy. The corrections, of which the chief is the thermal water equivalent of the apparatus, are not particularly difficult to determine. The skill and care required in the manipulation are not beyond any thoroughly expert operator, and the results are very satisfactory indeed, as may be seen by comparing those obtained by Berthelot and Stohmann in determining the heats of combustion of the same material in their respective laboratories.

The only drawback to the Berthelot bomb is the expense, as said above. This is due mainly to the amount of platinum used for the lining. The steel of which the bomb is made is especially exposed to corrosion. Berthelot protects the outside of the bomb by plating with nickel, which serves the purpose very well, as water and air are practically the only corrosive agents to which it is exposed. But with the interior the case is very different. The oxygen at high pressure is very active in itself. The carbon, sulphur, and phosphorus of the substances burned are completely oxidized, and carbonic, sulphuric, and phosphoric acids are formed. Indeed, Berthelot has shown that the apparatus may be used for determining carbon, sulphur, and phosphorus in these forms. More or less of the free nitrogen which is mingled with the oxygen used in the combustion is oxidized and forms nitric acid. It is necessary that the inner surface be covered with some substance which will resist these acids as well as the oxygen. Such materials are easily found, but the practical difficulty has been to find an inexpensive lining which will insure permanent protection to the steel. In Berthelot's first bomb, electroplating with platinum was tried, but the platinum soon began to scale off. After a few combustions gold was substituted for platinum and with better success, but it has not been used with oxygen under pressure.

MODIFICATIONS OF THE BOMB CALORIMETER BY MAHLER, HEMPEL, AND ATWATER.

Various modifications of Berthelot's apparatus have been devised especially to obviate the difficulty of expensive lining.

Mahler uses a bomb (fig. 12) of forged steel with enamel lining.[1] The cylinder is somewhat narrowed at the top and the cover is screwed directly upon it, the junction being made tight by a washer of lead. The enamel is easily put on or replaced, and it is stated that a single coating has been used for 300 combustions without injury. I have understood, however, that the enamel is apt to scale off in constant use. The form described by Mahler has an internal capacity of 600 cubic centimeters or nearly double that of Berthelot's bomb, as above described.

Hempel uses, for determinations of heats of combustion of coal, a simple bomb of steel without lining. This suffices for technical purposes, but is not recommended by him for scientific use.[2]

In accordance with suggestions by the author, Professor Hempel has most courteously had a bomb made by the mechanician who makes the bombs of his devising, and lined by Heraeus, of Hanau, with a thin sheet of platinum. The principle is the same as in Berthelot's form; but, whereas Berthelot's cover fits into the cylinder in the manner of a stopper, the cover in this rests directly upon the upper edge of the cylinder, a projection of the latter fitting into a groove in the former.

[1] Compt. rend., 113, 774, and 862, and Génie Civil, 1891, 20, No. 12, 198. See also Ztschr. analyt. Chem., 1893, 79, and Berthelot, Calorimétrie Chimique, 133.

[2] Hempel, Gasanalytische Methoden, 1890, 355. See also English translation; Methods of Gas Analysis, published by McMillan, New York.

A washer of lead is set in the groove of the cover and makes perfect closure feasible. The form of the apparatus is essentially the same as that shown in fig. 13. The quantity of platinum required is small, and the cost of the whole apparatus, including vise grip and lever for screwing the outer cover, a cylinder with compressed oxygen, fittings, manometer, and screw press for making hard pellets of the substance to be burned, was less than $220. It has served a most excellent purpose in our laboratory. Mr. C. D. Woods, who has done considerable work with it, has obtained results agreeing very closely with those of Berthelot and Stohmann.

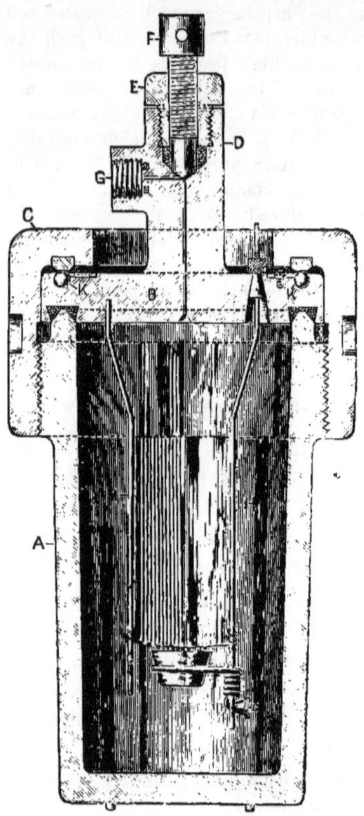

FIG. 13.—Bomb calorimeter as modified by Hempel and Atwater.

For measuring the temperature of the water, we use a thermometer made by Fuess and calibrated by the Physikalisch-technische Reichsanstalt in Berlin. It is graduated to hundredths of a degree, and the divisions are of such length as to permit easy estimation to thousandths by the aid of a magnifying glass. We are persuaded that in this way measurement of the temperature of water to thousandths of a degree can be made with as close an approach to accuracy as the weighings to the tenth of a milligram with an ordinary laboratory balance.

In place of copper for the cylinders outside the calorimeter cylinder, we have used vessels of "indurated fiber," which is a good nonconductor of heat. They have proved very satisfactory.

The bomb just described has not been found in every way satisfactory. It has, furthermore, seemed desirable to attempt to find a cheaper lining than platinum. Efforts are being made by the writer and his associate, Mr. Woods, to devise some needed improvements, find a less expensive lining than platinum, and develop an effective apparatus which may be made in this country at a cost that will bring it within the reach of ordinary laboratories. The results of these attempts are very encouraging, but are not yet ready for publication.

Figure 13, herewith, will help to explain the apparatus in the form in which we are using it. A represents the cylinder, C the screw cap, and B the cover of the bomb, which is placed upon the top of the cylinder and held down by the screw cap. A and C are made of gun steel. In a bomb lately made by the Pratt & Whitney Company, of Hartford, Conn., the steel is the same as is used for the Hotchkiss guns which are being manufactured by them. The metal has an unusually high tenacity and seems especially well fitted for the purpose. The cover (B) is provided with a neck into which fits a cylindrical screw (E), holding another screw (F). On the side of the neck is an aperture (G) between the lower end of D and the shoulder. In

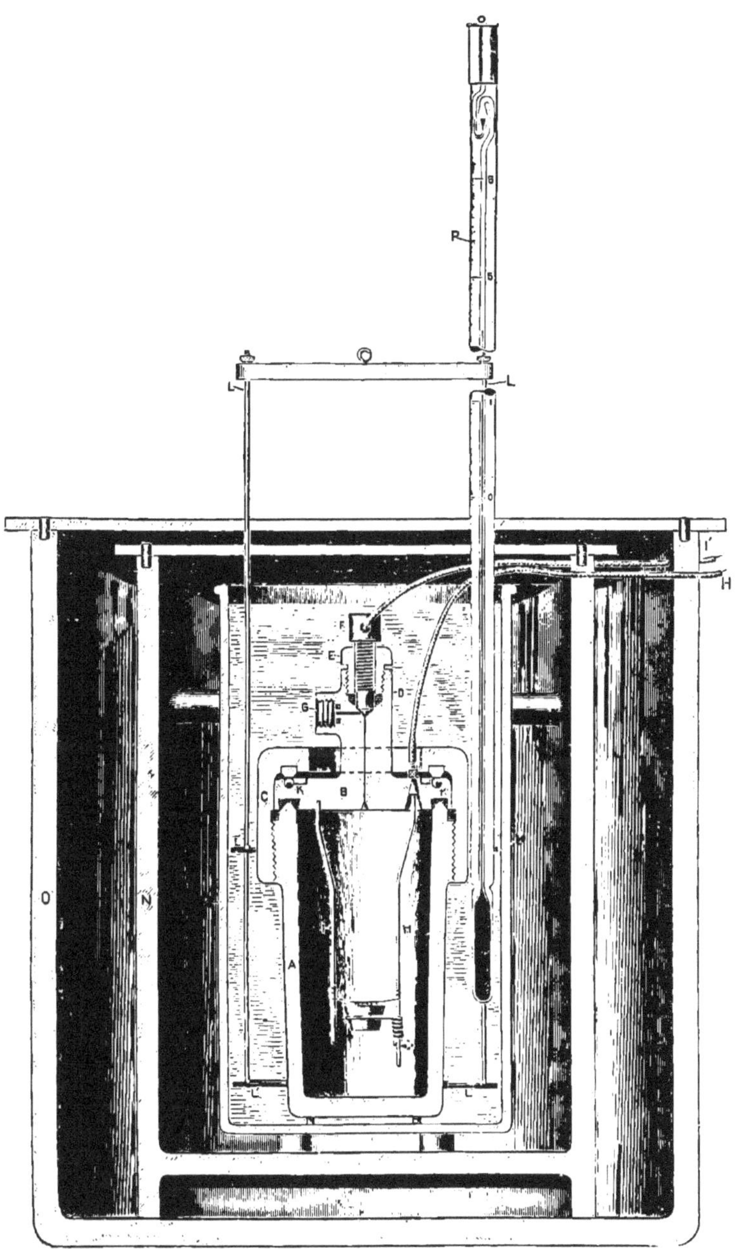

FIG. 14.—Bomb calorimeter of fig. 13, as mounted and standing in water.

D is a washer of lead on which the lower edge of E fits. By opening or closing the screw F the narrow passage from G is opened or closed. The opening is used for admitting oxygen at a high pressure through a narrow passage to charge the bomb. In B is an aperture through which passes a platinum wire (H), which is separated from the metal of the cover by insulating material. Hard vulcanized rubber serves very well for this purpose. Fastened to the lower side of the cover is another platinum rod (I), between which and H an electrical connection is made by a very fine iron wire. A screw ring holds the small platinum capsule, in which the substance to be burned is placed. At K K are ball bearings of hard steel to avoid friction in screwing the cap down.

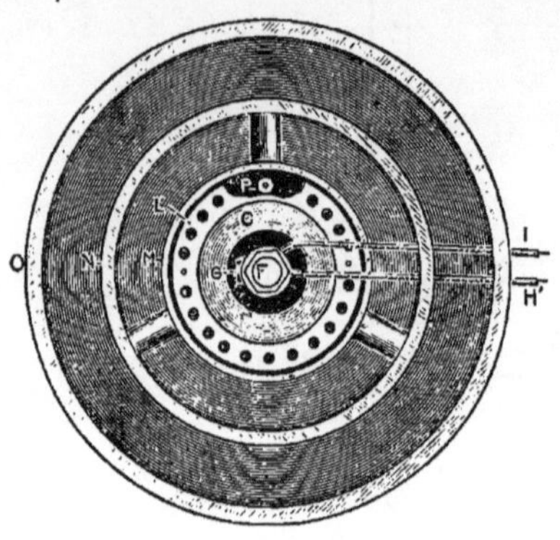

FIG. 15.—Top of bomb calorimeter of fig. 13.

Fig. 14 shows the bomb as mounted and standing in the water, which is contained in the calorimeter cylinder. The large cylinders (N and O) are made of indurated fiber and covered with plates of vulcanized rubber. A stirrer (L) serves for equalizing the temperature of the different portions of water after the combustion is completed. The situation of the thermometer (P) is likewise shown. Wires (H' I') serve for bringing the electrical current to the platinum wires in the bomb.

RESULTS OF DETERMINATIONS OF HEATS OF COMBUSTION.

The investigations by Frankland in 1866, above cited, like a great deal of the best pioneer work, were of the highest value for their purpose, but have not the accuracy which the later refinements of method have made possible. Determinations of the heats of combustions of a large number of substances by the Thompson-Stohmann method were made by Stohmann and his pupils, Von Rechenberg and B. Danilewski, between 1877 and 1885. Those of a smaller number were made by Rubner in 1885, and by Gibson in 1890, before the method of Berthelot became well known. The greater convenience, rapidity, and accuracy obtainable by use of compressed oxygen with the apparatus of Berthelot, the first description of which was published in 1885, led Stohmann, in 1887, to adopt this method. Since that time he has, with Langbein, not only reviewed his previous results, but added a large number of new ones by this method.

The following table summarizes the results obtained by use of the Thompson-Stohmann and Berthelot methods in determining the heats of combustion of the organic substances which occur in animals and plants, and are of special interest in physiological chemistry. They included all the determinations of this class which we have found up to July, 1894:[1]

[1] Since the above was written an extremely valuable article upon the subject by Professor Stohmann has appeared. See Experiment Station Record, vol. 6, and Ztschr. Biol., 31, 361.

Heats of combustion of organic substances.

	Berthelot method.		Thompson-Stohmann method.			
	Berthelot and associates.	Stohmann and Langbein.	Stohmann and associates.[1]	B. Danilewski.[2]	Rubner.[3]	Gibson.[4]
ALBUMINOIDS, ETC.						
Gluten	[5] 5,990.3			6,141		
Elastin		[6] 5,961.3				
Plant fibrin	[5] 5,832.3	[6] 5,941.6		6,231		
Serum albumin		[6] 5,917.8				
Syntonin		[6] 5,907.8				
Hemoglobin	[6] 5,910	[6] 5,885.1			5,950	
Milk casein	[5] 5,626.4	[6] 5,867	5,717	5,785		
Do		[6] 5,849.6				
Yolk of egg	8,112.4	[6] 5,840.9				
Legumin		[6] 5,793.1		5,573		
Vitellin	[5] 5,780.6	[6] 5,745.1				
Egg albumin	[5] 5,687.4	[6] 5,735.2	5,579			
Muscle, extractives and fat removed	[5] 5,728.4	[6] 5,720.5			5,778	
Crystallized albumin		[6] 5,672	5,598			
Muscle, fat removed		[6] 5,662.6	5,324		5,656	
Do		[6] 5,640.9				
Blood fibrin	[5] 5,529.1	[6] 5,637.1	5,511	5,709		
Harnack's albumen		[6] 5,553				
Wool	[5] 5,564.2	[6] 5,510.2				
Congluten		[6] 5,479	5,362			
Fibrin of skin		[6] 5,355.1				
Peptone		[6] 5,298.8		5,069		
Fish glue	[6] 5,240.1			5,493		
Chondrin	[5] 5,342.4	[6] 5,130.6		4,909		
Ossein	[5] 5,410.4	[6] 5,039.9				
Fibroin	[5] 5,095.7	[6] 4,979.6				
Chitin	[5] 4,655	[6] 4,650.3				
Tunicin	[5] 4,146.8					
Paraglobulin			5,637			
AMIDS, ETC.						
Urea	[7] 2,530.1	[6] 2,541.9	2,465	2,537	2,523	
Glycocoll	[8] 3,133.6	[6] 3,129.1	3,053			
Alanin	[8] 4,370.7	[6] 4,355.5				
Leucin	[8] 6,536.5	[6] 6,525.1				
Sarkosin		[6] 4,505.9				
Hippuric acid	[8] 5,659.3	[6] 5,668.2	5,642			
Aspartic acid	[8] 2,911.1	[6] 2,899				
Tyrosin	[8] 5,915.9					
Asparagin	[8] 3,396.8	[6] 3,514	3,428			
Kreatin (cryst)		[6] 3,714.1		[9] (3,206)		
Kreatin (water-free)		[6] 4,275.4				
Uric acid	[10] 2,754	[6] 2,749.9	2,621			
Guanin		[6] 3,891.7				
Caffein		[6] 5,231.4				
FATS.						
1. Animal:						
Fat of swine		[11] 9,476.9	9,380	[12] 9,686	9.423	9,515
Fat of oxen		[11] 9,485.7	9,357			9,427
Fat of sheep		[11] 9,493.6	9,406			9,530
Fat of horse			9,410			
Fat of dog			9,330			
Fat of goose			9,345			
Fat of duck			9,324			
Fat of man			9,398			
Butter fat		[11] 9,215.8	9,192			9,185
Sperm oil						10,001

[1] Jour. prak. Chem., 139 (N. F., 31), 273. These results were published in detail in Landw. Jahrb., XIII (1884), 413; later the figures were slightly changed as a result of the determination of the experimental corrections of the method and appear (corrected) as cited above (Jour. prak. Chem.).

[2] Centbl. med. Wiss., 1881, Nos. 26 and 27. Ref. Jahresb. Thier-Chem., 11 (1881), 7.

[3] Ztschr. Biol., 21 (1885), 250.

[4] Report of Storrs (Conn.) Experiment Station, 1890, 182.

[5] Ann. Chim. et Phys. (6), 22, 25.

[6] Jour. prak. Chem., 152 (N. F., 44), 1891, 336.

[7] Ann. Chim. et Phys. (6), 20, 13.

[8] Ibid. (6), 22, 1.

[9] Extract of meat.

[10] Compt. rend., 110, 1267.

[11] Jour. prak. Chem., 150 (N. F., 42), 1890, 361.

[12] Kind of fat not specified.

Heats of combustion of organic substances—Continued.

	Berthelot method.		Thompson-Stohmann method.			
	Berthelot and associates.	Stohmann and Langbein.	Stohmann and associates.[1]	B. Danilewski.[2]	Rubner.[3]	Gibson.[4]
FATS—continued.						
2. Vegetable:						
Olive oil (expressed)			9,328			9,471
Do			9,471			
Poppy-seed oil (expressed)			9,442			
Rape-seed oil (expressed)			9,489			
Do			9,619			
Ether extract of various seeds			{9,130 9,467}			
CARBOHYDRATES, ETC.						
1. Pentoses:						
Arabinose	[5]3,714	[6]3,722	3,695			
Xylose	[5]3,739.9	[6]3,746				
Fucose		[6]4,340.9				
Rhamnose (water-free)		[6]4,379.3				
Rhamnose (cryst)		[6]3,909.2				
2. Hexoses:						
Sorbinose		[6]3,714.5				
Galactose		[6]3,721.5	3,659			
Dextrose	[7]3,762	[6]3,742.6	3,692			3,754
Fructose		[6]3,755				
3. Heptoses:						
Glucoheptose	[8]3,732.8					
4. Disaccharids:						
Cane sugar	[9]3,961.7	[6]3,955.2	3,866		4,001	3,921
Milk sugar		[6]3,951.5	3,877			
Milk sugar (cryst)	[9]3,777.1	[6]3,736.8	3,663			3,710
Maltose		[6]3,949.3				
Maltose (cryst)		[6]3,721.8				
Trehalose		[6]3,947				
Trehalose (cryst)		[6]3,550.3				
5. Trisaccharids:						
Meletriose	[5]4,020	[6]4,020.8				
Meletriose (cryst)		[6]3,400.2				
Melezitose		[6]3,913.7				
6. Polysaccharids:						
Glycogen		4,190.6				
Cellulose	[10]4,200	[6]4,185.4	4,146			
Starch	[9]4,228	[6]4,182.5	4,123			4,164
Dextran	[9]4,180.4	[6]4,112.3				
Inulin	[9]4,187.1	[6]4,133.5	4,070			
ALCOHOLS.						
Ethyl alcohol	[11]7,068					
Glycerin		[12]4,112.4	4,317			
Mannite	[9]4,001.2	[6]3,997.8	3,908			3,950
Inosite	[5]3,676.8	[6]3,679.6				
ACIDS.						
Acetic	[11]3,490.4					
Palmitic		[13]9,352.9	9,226			
Stearic			9,429			
Oleic		[13]9,494.9				
Malonic	[14]1,998.2		1,960			
Succinic	[14]3,006.2		3,019			
Tartaric			1,745			
Citric	[14]2,477.9		2,397			

[1] Jour. prak. Chem., 139 (N. F., 31), 273. These results were published in detail in Landw. Jahrb. XIII (1884), 413; later the figures were slightly changed as a result of the determination of the experimental corrections of the method and appear (corrected) as cited above (Jour. prak. Chem.).
[2] Centbl. med. Wiss., 1881, Nos. 26 and 27. Ref. Jahresb. Thier-Chem., 11 (1881), 7.
[3] Ztschr. Biol., 21 (1885), 250.
[4] Report of Storrs (Conn.) Experiment Station, 1890, 182.
[5] Ann. Chim. et Phys. (6), 21, 409.
[6] Jour. prak. Chem., 153 (N. F., 45), 305, and private communication from Professor Stohmann.
[7] Ann. Chim. et Phys. (6), 13, 304.
[8] Compt. rend., 114, 921.
[9] Ann. Chim. et Phys. (6), 10, 455.
[10] Ibid. (6), 6, 546.
[11] Ibid. (6), 27, 310.
[12] Jour. prak. Chem. 150 (N. F., 42), 1890, 361.
[13] Ztschr. physikal. Chem., 10, 1892, 410.
[14] Ann. Chim. et Phys. (6), 23, 179.

NOTE.—The most of Berthelot's work is also referred to by Stohmann under references [5–15]. Under reference [16] will be found a table summarizing all of Stohmann's work with the bomb and some of the results obtained by Berthelot and his associates.

This field of inquiry is new and offers most excellent opportunity for useful work. The directions in which inquiry are now most needed are practically two—the study of compounds of interest in physiological chemistry and the study of the foods and feeding stuffs in which those compounds occur. In addition to the materials which are used in the nutrition of animals and plants the excretory products, which contain undigested residues and cleavage products from them, also demand investigation.

Heats of combustion of organic compounds.—The determination of heats of combustion of compounds of interest in physiological chemistry is desirable for two purposes, (1) for obtaining better knowledge of their chemical constitution, and (2) for securing the much-needed information as to their physiological uses, and especially their fuel values.

The indications which the heats of combustion give regarding the molecular constitution is of the greatest importance in the present condition of chemical science; and, aside from the purely theoretical interest of the inquiry there are various practical applications of value to the chemist; as, for instance, in distinguishing between isomeric compounds.

The importance of the measurements of fuel value of the compounds has already been indicated and hardly needs to be insisted upon further in this place.

The compounds which most demand study are of the kinds already studied and reported upon in the statements above. It is safe to say that all of the organic constituents of animal and vegetable tissues and their cleavage products, especially such as are formed by metabolism and living organisms, demand extended study; and it is of course desirable that the determinations of the heats of combustion should accompany and be supplementary to the investigation of their general characters, chemical and physical. In many laboratories these compounds are being isolated and studied. It is extremely desirable that the heats of combustion be determined at the same time.

What we need to do is to secure the materials in as large variety and of as great purity as possible and burn them in the calorimeter with due precaution, and record the results. If a number of investigators will undertake a research of this kind, the needed information will gradually accumulate.

Heats of combustion of foods and feeding stuffs, etc.—But the materials in which the compounds occur likewise demand calorimetric investigation. We may take, for instance, a food material, as wheat flour, study its composition, isolate its several constituents, and determine the potential energy of each, but at the same time it is desirable to burn the flour and observe whether its energy, as thus directly determined, corresponds with the energy as learned from the combustion of the several constituents. Investigations of this kind bring out discrepancies between the results obtained by combustion of the material as a whole and by separating the several ingredients and determining their heats of combustion and thus estimating the heat of combustion of the whole material. It is only by careful comparative studies of this kind that we shall learn the reasons for the discrepancies, the sources of error, and means for avoiding them, and the correct methods for determining the fuel values of food materials.

Although the work thus far done gives most cheering prospects of ultimate success, the results are not yet ripe for publication.

ISODYNAMIC VALUES OF NUTRIENTS.

It has been assumed that the heats of combustion of the several ingredients of food represent the potential energy which is transformed into kinetic energy when the food is used in nutrition. This principle may be expressed otherwise in saying that in their services as fuel to yield heat and mechanical (muscular) power their values will be directly proportional to the amounts of potential energy they contain. If this be true, we should expect that they would replace one another in proportion to their several heats of combustion. Late experiments, notably those of Rubner, imply that this is actually the case. A proper discussion of this subject will require the consideration of a large amount of detail. Reserving that for another occasion, I quote here a short popular description of experiments by Rubner which were made some ten years ago.[1]

Within a short time past, feeding trials with animals in the respiration apparatus have shown the proportions in which the several classes of nutritive ingredients of food do one another's work in serving as fuel in the body, and more extended experiments, with improved forms of the calorimeter, have given very accurate measurements of the amounts of potential energy in the same materials. The respiration experiments have been made with dogs, in the Physiological Institute in Munich, by Dr. Rubner, who has also made an extended series with the calorimeter. The largest number of the experiments with food materials in the calorimeter, however, have been conducted by Professor Stohmann, of the University of Leipsic, and his assistants. The results of experiments with the respiration apparatus and with the calorimeter agree with most remarkable closeness. In supplying the body with fuel, the protein, fats, and carbohydrates replaced each other in almost exact proportion to their heats of combustion. That the living body should thus be proved to use its food with such perfect chemical economy is certainly interesting and important. It is one more fact to add to the long lists that are bringing the functions of life more and more within the domain of ordinary physical and chemical law.

Isodynamic values for 100 parts of fat.

Nutritive substances, water free.	As determined by direct experiments with animals.	As determined by calorimeter.
Myosin	225	213
Lean meat	243	235
Starch	232	229
Cane sugar	234	235
Grape sugar	256	255

[1] Century Magazine, July, 1887.

The quantities of the several substances, lean meat, myosin (the chief protein compound of lean meat), starch, etc., are those which were found to yield the same amounts of heat when burned in the calorimeter, or to render the same service as fuel when consumed in the body of the animal, as 100 grams of fat. This explanation of the meaning of the expression "isodynamic values for 100 parts of fat" needs a little qualification to make it perfectly correct, but it is as accurate as I can well make it without going into a discussion too abstruse for the pages of a magazine, and it is really accurate enough for our purpose. The figures mean, then, that the dogs in the respiration apparatus obtained, on the average, as much heat to keep their bodies warm and energy for the work their muscles had to do, from 243 grams of lean meat) i. e., meat enough to furnish 243 grams of nutritive material after the water had been driven out), as they obtained from 100 grams of fat, while 235 grams of the lean meat, burned to equivalent products in the calorimeter, would yield the same amount of heat as the 100 grams of fat. Considering the great difficulties in experimenting with live animals, these two isodynamic values, 243 by the respiration apparatus and 235 by the calorimeter, agree very closely indeed. But with starch, the results by the two methods, 232 and 229, are still closer, while with ordinary table sugar and grape sugar they are as good as identical.

Taking our ordinary food materials as they come, and leaving out slight differences, due to the differences in digestibility, etc., Dr. Rubner has made the following general estimate of the amounts of energy in 1 gram of each of the three principal classes of nutrients:

Potential energy in nutrients of food.

	Calories.	Foot-tons.
In one gram of protein	4.1	6.3
In one gram of fats	9.3	14.2
In one gram of carbohydrates	4.1	6.3

These figures mean that when a gram (one-twentieth of an ounce) of fat, be it the fat of the food or body fat, is consumed in the body, it will, if its potential energy be all transformed into heat, yield enough to warm a kilogram of water 9.3 degrees of the centigrade theremometer, or, if it be transformed into mechanical energy such as the steam engine or the muscles use to do their work, it will furnish as much as would raise 1 ton 14.2 feet or 14.2 tons 1 foot. A gram of protein or carbohydrates would yield a little less than half as much energy as a gram of fat. In other words, when we compare the nutrients in respect to their fuel values, their capacities for yielding heat and mechanical power, an ounce of protein of lean meat or albumen of egg is just about equivalent to an ounce of sugar or starch; and a little over 2 ounces of either would be required to equal an ounce of the fat of meat or butter or body fat. The potential energy in the ounce of protein or carbohydrates would, if transformed into heat, suffice to raise the temperature of 113 pounds of water 1° F., while an ounce of fat, if completely burned in the body or in the calorimeter, would yield as much heat as would warm over twice that weight of water 1 degree.

One principle which they bring into clear relief is the remarkable economy with which the animal organism uses its material when the supply is limited, and the positive wastefulness it practices when the food supply exceeds the demand.

The dogs had very little room to move about inside the apparatus, and of course made very little muscular exertion. Hence they needed but little protein to make up for the wear of muscle; and, practically, the main demand of their bodies was for fuel to yield heat to warm their bodies and strength for the very little work their muscles had to do. When they fasted they consumed the fat and protein from the store in their bodies. How rigidly economical they were in this draft upon their previously accumulated capital was shown in the way that the consumption of fuel was affected by the temperature of the room. The interior temperature of the body remained very nearly the same, at "blood heat," all the while, as indeed it must, or the dogs would have died. In cold days more heat was radiated from the body than in warm, more was needed to supply its place, and more material was consumed. When the room was warmer the body burned less fuel; and the quantities consumed marked the changes of temperature with a delicacy almost comparable with that of the thermometer.

When the dogs had just food enough to supply their needs they used it with similar economy. In other words, when the income was equal to the necessary expenditure it was used as sparingly, as the sums taken from the capital had been. When the food supply was made larger, part of the extra material was stored in the body as fat and protein, but at the same time the daily consumption increased. That is to say, when their income was more liberal they laid part of it by, but at the same time allowed their current expenses to increase. It has been found by numerous experiments that when the nutrients are fed in large excess the body may continue for a time to store away part of the extra material, but after it has accumulated a certain amount it refuses to take on more, and the daily consumption equals the supply even when this involves great waste. With the large income the body continues for a time to add to its capital, but finally it comes to spend as much as it gets, and in so doing practically throws away what it can not profitably use.

Dr. Rubner's dogs showed in still another way their economy of fuel when the supply was limited, and wastefulness when they had more than they needed. The same animal that adjusted its consumption of fuel so accurately to the temperature of the air as long as the amount did not exceed its need, used it with no apparent regard to the temperature, whether warm or cold, as soon as the supply of food exceeded the necessary demand.

Physiologists have observed that the consumption of fuel in the body sometimes varies with the temperature and sometimes does not, and have been at a loss to explain the apparent discrepancies in their experimental results. These experiments help toward an explanation. But the interesting point is, not simply that the facts are learned, but that they are learned by studying the subject from the standpoint of the potential energy of the food. Previously the accounts have, so to speak, been drawn up in terms of protein, carbohydrates, and fats, and the balances have been difficult to calculate and still more difficult to explain. But in the experiments of which I have just been speaking, all the figures were reduced to terms of potential energy of the food and body substance consumed or stored. The results were calculated in calories, and the balancing of the accounts was thus made simple and the explanation plain.

Of course, I do not mean to say that we have thus suddenly come upon a complete explanation of the whole subject. This is simply an improvement of methods based on clearer understanding of principles and leading to clearer and more accurate results. It is, in short, the old story of clearing up an old mystery by use of a new and rational idea. As such, as well as for stronger reasons, it is of interest.

It is so easy to magnify the importance of any new discovery, and so hard to avoid going too far in drawing inferences from it, that I am inclined to put in another word of caution here. For instance, from the experiments above described one would infer that the food ingredients yield strength for muscular labor in exact proportion to their heats of combustion. But the dogs in the respiration apparatus performed no muscular work except that inside their bodies for respiration, keeping the blood in circulation, etc., and though we naturally assume that if they had used their muscles for exterior work, such as running or working a treadmill, the muscular energy yielded by the food would have been likewise equal to its potential energy, and though the other known facts make this assumption entirely probable, the experiments do not absolutely prove it. The production of muscular strength is a problem which is still but partly solved. Still I think it is reasonably safe to say that, in general, the foods that have the most potential energy are the ones that yield, not only the most heat to keep the body warm, but also the most strength for muscular work.

FUEL VALUES OF PROTEIN, FATS, AND CARBOHYDRATES.

The isodynamic values of the nutrients as computed by Rubner are for protein and carbohydrates each 4.1, and for fats 9.3 grams. It is hardly superfluous to repeat that these values are tentative, and that they can not apply with exactness to the nutrients of all food materials. They represent the results of the small number of experiments thus far made. Nearly all of these have been made with dogs, and the number of kinds of food material has been limited mainly to meats, a small number of kinds of fat, and to such carbohydrates as starch and sugar. In the estimates for protein, allowance is made for the nitrogenous material, chiefly urea, which does not undergo complete combustion in the body. The figures are based partly upon direct experiments and partly upon a priori estimates. Doubtless a more critical study of the subject will call for more or less revision of these figures for the fats that have been experimented with, and there is no question that different figures will have to be assigned to the nutrients of the same class from different foods and feeding stuffs. The correct estimates must be found by the same kinds of experimenting as are needed to confirm the theory that the law of conservation of energy actually applies in the living organism. These experiments will, of course, be calorimetric. They will be made in part by determinations of heats of combustion with the bomb calorimeter or other appropriate apparatus, but the chief dependence must be put upon experiments by which accurate determinations are made of the total income and outgo of the animal body as expressed in terms of both matter and energy. The respiration calorimeter is such an apparatus.

RESPIRATION CALORIMETERS.

During the past twelve years a number of efforts have been made to devise an apparatus by which accurate determination could be made of not only the chemical elements and compounds, but also the energy received and given off from the body. The most successful forms of which accounts have been published at all in detail are those of Rubner and Rosenthal. Each of these is essentially a small respiration apparatus, with a device for measuring the heat radiated from the body.

In the apparatus of Rubner[1] the respiration chamber consists essentially of a metal box with double walls. In the interior are arrangements for keeping an animal for a considerable length of time, as in an ordinary respiration apparatus. Provision is made for conducting a current of air through this chamber and for measuring its amount and determining its composition.

The apparatus of Rosenthal[2] is similar in principle to that of Rubner. The respiration chamber is double walled and the quantity of heat radiated from the animal is estimated by the expansion of the air

[1] Calorimetrische Methodik, Marburg, Elwert, 1891.

[2] Calorimetrische Untersuchungen, Arch. Anat. Physiol., Physiol. Abth., 1894, 223–282.

inclosed between the two walls. That is to say, in Rubner's apparatus the air, instead of being kept at a constant volume, is allowed to expand or contract, and the quantity of heat which passes through it is estimated from the increase of volume at constant pressure, while in the apparatus of Rosenthal the estimate is made by the variation of pressure at a constant volume. An especially noteworthy feature of the apparatus of Rosenthal is a device by which a definite volume of air is kept continually passing through the chamber. On emerging it passes through absorbents by which the respiratory products are removed, a fresh supply of oxygen being added at the same time, so as to maintain a constant volume. The space between the two walls of the box is filled with air, which is tightly confined so that none can escape. A manometer serves for determining the pressure, which is increased as the temperature rises and decreased as the temperature falls. The larger the amount of heat radiated from the body of the animal the higher will be the temperature, other conditions remaining the same, at which this confined portion of air is maintained, and hence the greater will be the pressure as registered by the manometer. Professor Rosenthal has been at no little pains to explain both the theory and the practice of this method of measuring the radiation of heat. A number of experiments have been made, but they are mostly of a character either special or preliminary, and in those thus far published no complete measurement of the income and outgo of material energy has been given. Rubner's arrangements for measuring and analyzing the air passing through the chamber are the same as those in the small respiration apparatus which was described above as devised by Voit on the principle of the apparatus of Pettenkofer. For full details of Rubner's apparatus as well as of Rosenthal's the original descriptions may be consulted.

RUBNER'S EXPERIMENTS WITH THE RESPIRATION CALORIMETER.

Rubner has lately published[1] the results of a very interesting series of experiments with dogs, in which the metabolism of material has been measured in terms of nitrogen and carbon, and that of energy in terms of the heat of combustion of the food and the heat radiated from the body. The animals were entirely at rest, and there was, therefore, no external work. Under these circumstances it was assumed that all of the energy given off from the body would be in the form of radiated heat. The number of these experiments has been rather small. The published account does not contain the full details. The results are in part estimated rather than directly determined, but, as given, they confirm both the correctness of the figures for isodynamic values of nutrients, which Rubner had previously annexed, and the application of the law of the conservation of energy in the organisms of the animals. The whole results may be summarized in the statement of

[1] De Quelle der thierischen Wärme, Ztschr. Biol., 30, 1893, 73.

Rubner to the effect that, considering the animal as a calorimeter, the heats of combustion of the nutrients are found to be the same as when determined by direct combustion with oxygen in the Thompson-Stohman, or bomb calorimeter.

OTHER RESPIRATION CALORIMETERS.

Professor Voit, of Munich, has devised a form of calorimeter for experiments with a man or other animal, but has, so far as I am aware, published no description, and the information which I have gained from brief personal inspection is hardly adequate or appropriate for publication.

Professor Chauveau, of Paris, has devised an apparatus which may be used for experiments with horses or oxen. It was in process of construction in the winter of 1892–93, but I have seen no detailed accounts of experiments made with it and can only refer to it in this very general way.

A small respiration calorimeter, on a plan based upon a principle very similar to that employed by Rosenthal and Rubner, has been lately described by Messrs. Haldane, White, and Washbourn.[1] Reference has also been made in print to an apparatus for similar purpose by Professor Burdon-Sanderson,[2] of Oxford; but I am not aware that any accounts of actual experiments with complete determinations of income and outgo of matter and energy by either of the instruments just named have been published.

An apparatus for similar purpose is now in process of development in the chemical laboratory of Wesleyan University. It is of a size appropriate to experiments with a man.

---

[1] Jour. of Physiol., 16, 1894, 123.

[2] Phil. Mag. [5], 29, 306.

CHAPTER VIII.

PECUNIARY ECONOMY OF FOOD.

The cost of food is the principal item of the living expenses of the majority of people—of all, indeed, but the especially well-to-do in Connecticut and the other Eastern States.[1] In the report of the Bureau of Statistics of Labor of Massachusetts for 1884 are summarized the results of investigations into the cost of living of people with different incomes in Massachusetts, in Great Britain, and in Germany. Dividing expenses into those for food, clothing, rent, fuel, and sundries, the percentage of the whole income expended for food averages as follows:

Percentage of family income expended for subsistence.

	Annual income.	Expended for food.
GERMANY.		*Per cent.*
Workingmen	$225 to $300	62
Intermediate class	450 to 600	55
In easy circumstances	750 to 1,100	50
GREAT BRITAIN.		
Workingmen	500	51
MASSACHUSETTS.		
Workingmen	350 to 400	64
Do	450 to 600	63
Do	600 to 750	60
Do	750 to 1,200	56
Do	Above 1,200	51

The large majority of families in this country are said to have not over $500 a year to live upon. More than half of this goes, and must go, for food. The cost of preparing food for the table, rent, clothing, and all other expenses must be provided from the remainder.

These statements apply less accurately to farmers than to the inhabitants of the larger towns, but, although the farmer produces much of his food, yet taking everything into account the expense for nutriment is large even for him.

Late statistics published by the United States Department of Labor imply a smaller relative cost of food in the Southern and Western States, where food materials are cheaper, than in the Eastern States. In some cases the expense falls below 50 per cent of the total earnings of wage workers.

Although the cost of food makes so large a part of the whole cost of living, and although the health and strength of all are so intimately connected with and dependent upon their diet, yet even the most intelligent people know less of the actual uses and values of their food for fulfilling its purposes than of almost any other of the necessities of life.

[1] In portions of the West and South, where food is less expensive, its cost is somewhat less in comparison with other living expenses.

CHEAP V. DEAR FOOD.

The cheapest food is that which supplies the most nutriment for the least money. The most economical food is that which is the cheapest and at the same time best adapted to the wants of the user. The maxim that "the best is the cheapest," does not apply to food. The best food, in the sense of that which is the finest in appearance and flavor, and which is sold at the highest price, is not generally the cheapest, nor is it always the most healthful or economical.

Of the different food materials which the market affords, and which are palatable, nutritious, and otherwise fit for nourishment, what ones are peculiarly the most economical? There are various ways of comparing food materials with respect to the relative cheapness or dearness of their nutritive ingredients. One, and perhaps the best, consists in comparing the nutrients obtained for a given sum in different materials. Table 16, which follows, gives estimates of amounts of nutrients that could be purchased for 25 cents at the rates named. The calculations are based upon the analyses in Table 1 and upon the retail prices current in Connecticut.

The figures of the table tell their story so plainly that they need very little comment. A quarter of a dollar invested in the sirloin of beef at 22 cents per pound pays for 1¼ pounds of the meat with three-eighths of a pound of actually nutritive material. This would contain one-sixth of a pound of protein and one-fifth of a pound of fat, and supply 1,120 calories of energy. The same amount of money paid for oysters at the rate of 50 cents per quart brings 2 ounces of actual nutrients, an ounce of protein, and 230 calories of energy. But in buying wheat flour at $7 a barrel the 25 cents pays for 6¼ pounds of nutrients with eight-tenths of a pound of protein and 11,755 calories of energy.

The price of food is not regulated solely by its value for nutriment. Its agreeableness to the palate or to the buyer's fancy makes a large factor of the current demand and market price. There is no more nutriment in an ounce of protein or fat of the tenderloin of beef than in that of the round or shoulder. The protein of animal foods does, however, have an advantage over that of vegetable foods. Animal foods, such as meats, fish, milk, and the like, gratify the palate in ways which most vegetable foods do not, and, what is perhaps of still greater weight in regulating the actual usage of communities by whose demand the prices are regulated, they satisfy a real need by supplying protein and fats, which vegetable foods lack.

People who can afford it, the world over, will have animal foods and will compete with one another in the prices they give for them. In general, the animal foods are more easily and completely digested than vegetable. There is doubtless good ground for paying somewhat more for the same quantity of nutritive material in the animal food.

For persons in good health the foods in which the nutrients are most expensive are like costly articles of adornment. People who can well afford them may be justified in buying them, but they are not economical.

CHART 2.—PECUNIARY ECONOMY OF FOOD.

Amounts of actually nutritive ingredients obtained in different food materials for 25 cents.

[Amounts of nutrients in pounds. Fuel value in calories.]

Protein. Fats. Carbohydrates. Fuel value.

	Price per pound.	Food materials for 25 cents.
	Cts.	*Lbs.*
Beef, sirloin	25.0	1.00
Beef, round	15.0	1.67
Beef, neck	6.0	4.17
Mutton, leg	22.0	1.14
Ham, smoked	16.0	1.56
Salt pork, very fat	12.0	2.08
Codfish, fresh	8.0	3.13
Codfish, salt	7.0	3.57
Mackerel, salt	12.0	2.08
Oysters, 35 cents quart	18.0	1.43
Eggs, 25 cents dozen	14.7	1.70
Milk, 7 cents quart	3.5	7.14
Cheese, whole milk	15.0	1.67
Cheese, skim milk	8.0	3.13
Butter	30.0	0.83
Sugar	5.0	5.00
Wheat flour	3.0	8.33
Wheat bread	7.0	3.57
Corn meal	2.5	10.00
Beans	5.0	5.00
Potatoes	1.2	20.00
Standard for daily diet for man at moderate work	German.*	
	American.†	

Weights of nutrients and calories of energy in 25 cents' worth

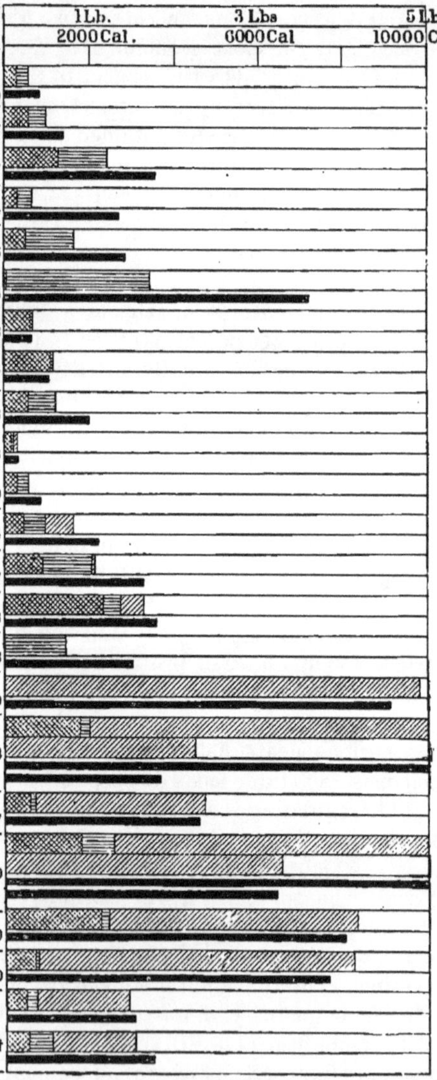

* Voit. † Atwater.

TABLE 16.—*Amounts of nutrients furnished for 25 cents in food materials at prices in Eastern States.*

Food materials as purchased.	Prices per pound.	25 cents will pay for— Total food materials.	Nutrients. Total.	Protein.	Fats.	Carbohydrates.	Calories of potential energy.
MEATS, ETC.	*Cents.*	*Pounds.*	*Pounds.*	*Pounds.*	*Pounds.*	*Pounds.*	*Calories.*
Beef, neck	8	3.13	0.95	0.49	0.44	....	2,765
	6	4.17	1.27	.65	.58	....	3,655
Chuck ribs	16	1.56	.56	.23	.31	....	1,735
	12	2.08	.75	.31	.42	....	2,350
Ribs	22	1.14	.47	.14	.32	....	1,610
	18	1.39	.57	.17	.39	....	1,960
Shoulder	14	1.79	.57	.30	.25	....	1,615
	10	2.50	.79	.43	.34	....	2,235
Sirloin	22	1.14	.37	.17	.19	....	1,120
	18	1.39	.45	.21	.23	....	1,360
Rump	18	1.39	.63	.19	.43	....	2,170
	15	1.67	.76	.23	.52	....	2,620
Round:							
First cut	18	1.39	.44	.25	.17	....	1,180
	15	1.67	.52	.30	.21	....	1,445
Second cut	10	2.50	.52	.35	.15	....	1,285
	8	3.13	.65	.44	.18	....	1,580
Flank, corned	15	1.67	.74	.21	.49	....	2,460
	10	2.50	1.11	.31	.73	....	3,655
Corned and canned	18	1.39	.66	.37	.24	....	1,700
	14	1.79	.85	.48	.31	....	2,200
Liver	8	3.13	.96	.63	.17	.11	2,095
Mutton:							
Shoulder	20	1.25	.44	.19	.24	....	1,365
	15	1.67	.58	.25	.31	....	1,775
Leg	25	1.00	.31	.15	.16	....	955
	20	1.25	.39	.19	.20	....	1,195
Loin	25	1	.43	.13	.29	....	1,465
	20	1.25	.53	.16	.37	....	1,860
Pork, rib roast	12	2.08	.88	.28	.58	....	2,970
	10	2.50	1.06	.34	.70	....	3,585
Smoked ham, whole	16	1.56	.81	.23	.54	....	2,705
	12	2.08	1.08	.31	.72	....	3,615
Salt fat pork	15	1.67	1.46	.02	1.38	....	5,860
	12	2.08	1.83	.02	1.72	....	7,295
Pork, sausage	15	1.67	.98	.23	.72	....	3,465
	12	2.08	1.22	.29	.89	....	4,295
Poultry, etc., chicken	22	1.14	.20	.17	.02	....	400
	16	1.56	.27	.24	.02	....	530
Turkey	23	1.09	.25	.18	.06	....	590
	18	1.39	.31	.22	.08	....	745
Fish, shad, whole	15	1.67	.25	.15	.09	....	660
	10	2.50	.37	.23	.12	....	935
Mackerel, whole	18	1.39	.22	.14	.06	....	515
	15	1.67	.25	.17	.07	....	610
	10	2.50	.37	.25	.11	....	930
Bluefish, dressed	15	1.67	.19	.16	.01	....	340
	10	2.50	.28	.25	.02	....	550
Striped bass, whole	18	1.39	.14	.12	.01	....	265
	12	2.08	.21	.17	.02	....	400
Haddock, dressed	8	3.13	.28	.26	.01	....	525
	5	5	.45	.41	.01	....	805
Cod, dressed	10	2.50	.29	.27	.01	....	545
	8	3.13	.36	.33	.01	....	655
	6	4.17	.48	.44	.01	....	860
Halibut, steaks	20	1.25	.26	.19	.06	....	605
	16	1.56	.32	.24	.07	....	740
Salt cod	8	3.13	.55	.50	.01	....	970
	5	5	.85	.80	.02	....	1,570
Salt mackerel	16	1.56	.49	.23	.24	....	1,440
	12	2.08	.66	.31	.32	....	1,925
Canned salmon	20	1.25	.46	.25	.20	....	1,310
Oysters:							
50 cents per quart	25	1	.13	.06	.01	.04	230
35 cents per quart	17.5	1.43	.18	.09	.02	.05	345
Lobsters:							
Whole	12	2.08	.14	.11	.01	....	245
Canned	10	2.50	.17	.14	.02	....	345
	20	1.25	.28	.23	.01	....	470
EGGS AND DAIRY PRODUCTS.							
Eggs:							
35 cents per dozen	25	1	.23	.12	.10	....	645
25 cents per dozen	18.2	1.37	.32	.17	.14	....	910
15 cents per dozen	11	2.27	.53	.28	.23	....	1,490

TABLE 16.—*Amounts of nutrients furnished for 25 cents in food materials at prices in Eastern States*—Continued.

Food materials as purchased.	Prices per pound.	25 cents will pay for—					
		Total food materials.	Nutrients.				Calories of potential energy.
			Total.	Protein.	Fats.	Carbohydrates.	
MISCELLANEOUS.							
Milk:	*Cents.*	*Pounds.*	*Pounds.*	*Pounds.*	*Pounds.*	*Pounds.*	*Calories.*
8 cents per quart	4	6.25	.81	.23	.25	.29	2,020
6 cents per quart	3	8.33	1.08	.30	.33	.39	2,675
4 cents per quart	2	12.50	1.63	.45	.50	.59	4,045
Butter	35	.71	.64	.01	.60	....	2,550
	30	.83	.74	.01	.71	....	3,015
	25	1	.90	.01	.85	.01	3,625
Cheese, whole milk	18	1.39	.96	.40	.49	.02	2,850
	15	1.67	1.17	.47	.59	.03	3,420
	12	2.08	1.45	.59	.72	.04	4,210
Potatoes:							
$1 per bushel	1.67	15	2.69	.27	.01	2.28	4,785
80 cents per bushel	1.33	19	3.41	.34	.02	2.09	6,090
50 cents per bushel	.83	30	5.38	.54	.02	4.56	9,570
Sweet potatoes	5	5	1.27	.07	.02	1.14	2,335
	3	8.33	2.11	.10	.03	1.89	3,830
Beets	2	12.50	1.23	.16	.01	.94	2,090
	1	25	2.45	.32	.02	1.87	4,180
Turnips	2	12.50	.93	.11	.02	.72	1,630
	1	25	1.86	.22	.04	1.44	3,260
Sugar	5	5	4.90	....	....	4.89	9,095
	6	4.17	3.64	.96	.09	2.47	6,760
Dried beans	5	5	4.37	1.15	.10	2.96	8,065
	4	6.25	5.46	1.44	.13	3.70	11,110
Maize, corn meal	3	8.33	7.08	.77	.32	5.88	13,720
	1	25	21.25	2.30	.95	17.65	41,160
Oatmeal	5	5	4.62	.76	.30	3.41	9,275
	4	6.25	5.47	.69	.07	4.68	10 285
Wheat flour	3.5	7.14	6.25	.79	.08	5.35	11,755
	3	8.33	7.29	.92	.09	6.24	13,695
Wheat bread	7	3.57	2.42	.31	.06	2.01	4,570
	5	5	3.38	.44	.09	2.62	6,445
Crackers, Boston	12	2.08	1 91	.22	.21	1.43	3,955
	8	3.13	2.87	.33	.31	2.15	5,920

CHAPTER IX.

FOOD CONSUMPTION.

A most important branch of the subject here considered is the actual food consumption of people of different countries and classes.

STUDIES OF DIETARIES—HISTORICAL SUMMARY.

A large number of observations have been made in Europe to learn the amounts of food and of actual nutrients consumed by people of different classes and occupations and under different circumstances. Within a very few years past like studies of dietaries have been undertaken in the United States and in Japan.

The method of making such observations consists in finding the amounts of food materials of different kinds consumed by one or more persons during a certain number of days and calculating the quantities of nutrients in each material from its amount and composition.

During some time past I have, with the aid of Drs. H. B. Gibson and C. F. Langworthy, endeavored to collate the results of the studies made in this direction up to the present time. We have found records of the examination of 491 separate dietaries, exclusive of army rations. The earliest were made in 1851 in England by Beneke, a German physician, afterwards professor in the University of Marburg. The majority have been made during the past 15, and by far the larger number of the most reliable ones during the past 10, years. The people whose dietaries have been studied have been of various classes, ages, and occupations. A few were in professional life and were decidedly well to do. The most were wage workers. Some of these were very poor, but the larger number were in reasonably comfortable circumstances as compared with the majority of people of like occupation in the countries where they lived.

Some of the studies were made with the greatest care and thoroughness, the food was accurately weighed and analyzed, and strict account was kept of the number, sex, and occupation of the persons who were nourished by it. In a few instances pains were taken to determine the proportions actually digested. But in a majority of the dietaries reported upon the amounts and composition of the food instead of being determined exactly by weighings and analyses were more or less roughly estimated, so that the results lack scientific accuracy.

From the 491 dietaries we have selected 338 as accurate enough to warrant their use in drawing inferences. The number of persons whose food consumption was observed in each dietary of this selected list varied from a single individual to several hundred, and the time of observation in each case from 1 to 30 or more days. In addition to these the studies of some 41 Japanese and other Asiatic dietaries have been lately reported by Mori, Taniguti, and Eijkman, but we have not access to the details, a circumstance which is the more to be regretted because of the high value of the work.

In the selected list of 338 (columns A and B of table herewith) are included 101 studies of dietaries in the United States and Canada, of which 38 were in New England, 25 in Philadelphia, and 26 in Chicago. This list of 338 we have divided in two classes. The first including all the studies that seem to us reasonably accurate and complete; the second including those which are less accurate but sufficiently so to allow their results to be included in the general averages. The classification of all these dietaries by countries and by completeness of detail is summarized in the following tabular statement:

Number of dietaries collated, classified by countries and by completeness of detail.

[A. Reasonably accurate and complete. B. Less accurate, but included in averages. C. Not included in averages.]

	A	B	C	A+B	A+B +C
Europe:					
England	......	7	49	7	56
Scotland	......	......	6	......	6
Wales	......	......	1	......	1
Ireland	......	......	2	......	2
Total Great Britain	......	7	58	7	65
France	......	1	5	1	6
Belgium	......	......	15	......	15
Denmark	2	......	......	2	2
Sweden	13	......	......	13	13
Russia	4	25	......	29	29
Bavaria	21	34	7	55	62
Saxony	......	40	3	40	43
Prussia	3	13	15	16	31
Other German States	30	10	10	40	50
Total Germany	54	97	35	151	186
Austria	......	3	......	3	3
Switzerland	2	......	10	2	12
Italy	15	......	15	15	30
Spain	......	......	5	......	5
Portugal	......	......	1	......	1
Total Europe	90	133	144	223	367
Asia:					
India	......	......	7	......	7
Java[1]	1	......	......	1	1
Japan	8	5	1	13	14
Total Asia	9	5	8	14	22
Canada	......	13	......	13	13
United States:					
Massachusetts	......	19	......	19	19
Connecticut	13	5	1	18	19
Pennsylvania, Philadelphia	......	25	......	25	25
Illinois, Chicago	......	26	......	26	26
Total United States	13	75	1	88	89
Total North American	13	88	1	101	102
Grand total	112	226	153	338	491

[1] Java Village, World's Fair, Chicago.

All of these studies in the United States were observed during the past 8 years. Those in Philadelphia and Chicago were observed by Miss Amelia B. Shapleigh in the use of the Dutton Fellowship of the College Settlements Association, 1892–93. The rest were studied by the writer and his associates at Wesleyan University, in cooperation with the Massachusetts Bureau of Labor and the United States Department of Labor, and as part of the work of the Storrs (Conn.) Experiment Station.

Of the European dietaries in the list only 7 are from England; these were estimated by Playfair some 30 years ago. Only 1 comes from France. Fifteen are from Sweden and Denmark and 29 from Russia. From Germany there are 151, of which the earliest were by Liebig and the larger number are by Voit and his followers. From Italy are 15, including 8 by Manfredi and 6 by Albertoni and Mori, which are among the latest and most thoroughly studied of all. From Japan are 13, all of which have been made lately by Germans connected with the University of Tokyo and by Japanese working with them. One was the dietary of Javanese in the Java Village at the World's Fair. This was studied in connection with the examinations of foods at the fair, under the direction of the writer, which were referred to in the statements regarding analyses of foods above.[1]

The data thus collected are far from sufficient for satisfactory conclusions. Indeed, perhaps the most important lesson they teach is the need of more such studies. Their general character may be inferred from the selections in Table 37.

The only European country from which the studies are numerous and accurate enough to be taken as in any way representative of the national food consumption is Germany, although those from Denmark, Russia, and Italy are most interesting and instructive. The American data are confined to Massachusetts, Connecticut, Philadelphia, Chicago, and a few places in Canada. The right of those from the Eastern States to be taken as representative is confirmed, in a general way, by the figures for food consumption in the family budgets published by the Massachusetts bureau and the United States Department of Labor, especially in late reports of the latter on the cost of production.

SPECIAL INVESTIGATIONS OF DIETARIES.

It will be to the purpose here to cite some of the special studies of dietaries that have been made up to the present time as illustrative of the character of the work done, the methods followed, and the excellencies which may be imitated and the defects which should be abolished in future inquiry in this most important direction. To this end I select:

(1) Several studies of dietaries in the United States, including some

---

[1] Page 20.

of the earliest and the latest. The former are the least accurate; the latter are more thorough, though they are very far from being all that could be desired.

(2) A series made in Germany, which show how great care may be used in such inquiry and at the same time serve to illustrate the possibility of further improvement of method.

(3) A series of studies lately made in Italy, in which a decided advance in method is made in certain directions.

These illustrations will be followed by a résumé of the work in this direction up to the present time.

DIETARY STUDIES IN THE UNITED STATES.

In connection with the Massachusetts Bureau of Statistics of Labor, a series of studies of dietaries of factory operatives, mechanics, and other people with moderate incomes, in private families and boarding houses, were made by the writer in 1886. At the same time, and later, several dietaries of students and laborers and one of a well-to-do private family in Middletown, were examined. These were the first at all extended inquiries in this direction with which I am familiar in the United States. Although the work thus done represented only the beginning of an investigation of an important subject, the result seemed to warrant, or at least to suggest, generalizations of no little interest, and at the same time served to indicate directions in which further inquiry is needed. The following account of the Massachusetts and Canadian studies is taken from a report by myself and Mr. C. D. Woods in the report of the Storrs (Conn.) Experiment Station for 1891.

INVESTIGATION OF DIETARIES IN CONNECTION WITH THE MASSACHUSETTS BUREAU OF STATISTICS OF LABOR.

In order to obtain some definite information in regard to the ways of living and especially the food of factory operatives, mechanics, and other working people of native and foreign birth in Massachusetts, the statistics of the amounts and costs of food consumed, and age, sex, and occupation of the consumers, were collected by the bureau under the direction of its chief, Hon. Carroll D. Wright, in several manufacturing cities in the State; and as many of the people, whose conditions of life and labor were thus studied, were French Canadians, who come in large numbers to Massachusetts, as to other States bordering upon Canada, and form a not unimportant factor of the population, an agent of the bureau visited Canada and collated similar statistics regarding the people in the places from which they come. The data thus collated were placed in my hands, and with the assistance of Mr. E. W. Rockwood, then assistant in the chemical laboratory of Wesleyan University, the quantities of nutritive ingredients were estimated, the results of analyses of food materials to those of the dietaries being used as the basis of the calculations.

From a larger number collated by the bureau, 30, which were regarded as representative, were selected for chemical examinations and the making up of average results. The 30 dietaries summarized in the tables of recapitulations are divided in three series, as follows:

Classification.	Series A, miscellaneous, Massachusetts.	Series B, French Canadian, Massachusetts.	Series C, French Canadian, Canada.
Data given in tables of recapitulation	5	5	5
Data not given, but results included in averages in tables of recapitulation	[1]5	[1]2	8
	10	7	13

[1] Series A included 3 French Canadian dietaries which are averaged with those of series B, making 10 of the latter, all told, in the averages of series B, and only 7 in the averages for series A in the tables of recapitulations, beyond.

A. *Miscellaneous, Massachusetts.*—These include 10 dietaries of families and boarding houses. The families are nearly all laboring people, while the boarders in the boarding houses are mostly operatives in mills and factories, though some are clerks, dressmakers, etc. A few are French Canadians.

The 5 dietaries of this series, of which the data are given in the tables just referred to, include 3 of boarding houses and 2 of families in Lowell, Lynn, East Cambridge, and Boston. Two more of boarding houses, 1 in Lowell and 1 in Lawrence, are included only in the averages, for this series, in the tables. The results for 3 dietaries of French Canadian families in North Cambridge, of this series, are included in the averages for series B in the tables of recapitulations. The persons are factory and mill operatives, mechanics, etc., with a few clerks and dressmakers.

B. *French Canadians, Massachusetts.*—These include 7 dietaries of families and boarding houses, all of French Canadians. The 5 dietaries of this series which are used in the tables include those of 3 families and 2 boarding houses in Holyoke, Lawrence, and Lowell. Of those included only in the averages, 2 were of families in Worcester, of this series, and 3 of families in East Cambridge, of series A. With the exception of women, children, and others engaged in household duties, or in no actual labor, the people are mostly mill and factory operatives; a few are brickmakers.

C. *French Canadians, Canada.*—These include dietaries of 13 families and boarding houses in Montreal, Quebec, and other places in Canada. The people are represented as all belonging to the laboring classes. The 5 dietaries of this series which are used in the tables include 1 of a boarding house and 4 of families in Quebec, St. John, Sherbrooke, Richmond, and Rivière du Loup. The averages include, with these, 8 others of families and boarding houses.

Data.—In the descriptions given in the report of the kinds of data employed and the ways in which they were attained, they were classified as follows:

Class A. Those contained in statistics as collected by the agents of the bureau. They have to do with:

I. Statistics of food materials, including: *a*, kind; *b*, quantity; *c*, costs.

II. Statistics of consumption of food materials, including: *a*, number; *b*, sex; *c*, age; *d*, occupation of persons nourished, and *e*, time.

Class B. Data obtained from other sources and used in the computations. They have to do with:

I. Chemical composition of the food materials. Proportions of nutrients (nutritive ingredients) in each. These figures were employed in computing the quantities of actual nutrients in the dietaries.

II. The proportions of nutrients required by persons differing in age, sex, etc. These proportions were used in estimating the numbers of men at moderate work who would be equivalent in demand for food to the men, women, and children nourished by the food for each dietary, the object being to place all the dietaries on a uniform basis for comparison.

In collecting the statistics of Class A, the agents of the bureau visited the houses where the people lived, and the cotton mills, shoe factories, glass factories, machine shops, blacksmith shops, brickyards, and other establishments where they worked, examined the bills of dealers for food furnished, the family and boarding house accounts, and by these and other appropriate means secured as accurate figures as practicable.

The chief source of error is undoubtedly to be found in the fact that the statistics give the amount of food purchased, not that actually eaten. How much was thrown away as refuse or otherwise wasted can not be ascertained.

Of the data of Class B, those regarding the composition of the food materials used were estimates rather than the record of actual analysis of the materials which would be required for entire accuracy. For the estimates the results of analyses made in this laboratory, and included with those reported in the chapter on "Composition of food materials," were employed. The other American analyses then available, mostly of dairy and cereal products, were also used. For the few materials of which no American analyses had been made, recourse was had to European figures. The data and methods used in estimating the composition of the food materials are given in detail in the report. The following quotations will give an idea of what they were. The tables referred to are too extensive to be inserted here. The analyses of beef were those described in the chapter on "Composition of food materials" above, as made for the United States National Museum.

The method of estimating the composition of beef was as follows:

A large amount, the larger part, we are informed, of the beef consumed in many of our Eastern cities is so-called "Chicago" or "Western" beef, which is slaughtered in Chicago or elsewhere and brought East. From a car load of Chicago beef a side was selected by an experienced dealer as of average quality, especial pains being taken to secure one of average fatness. This side of beef was divided into 25 pieces, or "cuts," in the manner common in New York markets, and portions of each piece, sufficient to represent the whole, were analyzed, the proportions of refuse (bone, gristle, etc.), water, and nutrients being determined.

A diagram representing these divisions of the beef was placed in the hands of the collectors of the dietaries here examined, who, so far as practicable, indicated in their statements the parts of the animal from which the beef of the several dietaries was taken. The manner of cutting up the beef differs in different places, but not sufficiently to very materially affect the estimates. A more serious matter is the variation of different specimens of beef, and it is, of course, a question how close the side selected for analysis, as above stated, comes to representing the average of the kinds in the dietaries. We are informed that in all the Massachusetts cities, where the dietaries were collected, nearly all the beef used is so-called Chicago beef, and it is probable that the analyses fairly indicate the quality of the beef sold.

As the best way for utilizing these data, an assistant has gone over the dietaries, noted the cuts of beef where stated, and ascribed to each the percentages of nutrients found in the analyses of corresponding cuts. The results are shown in the following table. Where the original includes two or more cuts in one entry, the average is taken. The several computations for roast beef are averaged together. The same is done for beef stew, beefsteak, etc., and some of these latter averages are incorporated in the table beyond, giving the percentages of nutrients in food materials assumed in analyses of dietaries.

The figures used in calculating the amounts of nutrients in the dietaries are generally given in the table showing the percentages of nutrients in food materials assumed in analyses of dietaries. In some special cases, however, they are not given in this table, but are stated with explanations in the explanatory notes appended to the details of the dietaries in which they are used.

To insure perfect accuracy it would, of course, be necessary to analyze the materials actually used in each case. It is probable that while divergences, in some cases very wide, might occur, the figures for the composition of each dietary, as a whole, would be substantially accurate.

The item about which there seems to be the most question is the quantity of fat in the meats, especially the beef. The analyses here used accord very closely with European figures for very fat beef.[1] Numerous observations, however, which can not be detailed here, but which seem to be but little short of decisive, imply that the beef commonly used on the Continent of Europe is, on the average, less fat than the average beef in our markets. It is certain that much of that commonly used in our Eastern cities is very much fatter than that here analyzed and taken as the basis of these computations.

Attention has been called elsewhere to the fact that the figures for weights of food materials in the dietaries represent the quantities purchased and do not indicate how much was eaten. The rejection of a considerable part of the fat of meats by many persons is one of the most common of dietary facts, at least in the Northern and Eastern States. Some of the fat of beef is left with the butcher, much goes to the soap maker, and much more into the garbage. But a surprisingly large part of the fat of our beef is so diffused through the lean, much of it in invisible particles, that when we have cut out the larger pieces of fat from our roast beef or our steak and left them on our plates, we, nevertheless, eat the bulk of the actual fat of the meat with the lean and the small portions of visible fat which adhere to it.

Especial stress is laid on this point, because the dietaries here studied indicate a remarkably large consumption of fat in this country. The possible bearing of this fact upon our national dietetics may be extremely important.

---

[1] See analyses quoted by König, Nahrungsmittel, Bd. I.

In the statistics of some of the Massachusetts dietaries, and most of those collected in Canada, the quantities of certain food materials, especially vegetables, were not stated, and had to be estimated from the costs. The uncertainty as to the accuracy of the total estimates was thus increased, though, as was believed, not enough to seriously affect the value of the results for the purpose for which they were intended, namely, an exhibit of the general character of the food consumption and a means toward learning how such an inquiry may be best conducted.

The method for estimating the number of men, at moderate work, who would be equivalent in demands for nutriment to the persons partaking of the food of each dietary, is thus explained in the report:

Since the people nourished by the dietaries here examined differ in age, sex, and occupation, and hence differ likewise in their demands for nutriment, and since a chief object of the examination is to compare the dietaries with one another in respect to the quantities of actual nutrients supplied, it is clear that to attain our object we need some standard for estimating the relative demands of people of different classes. If, for instance, we could take a particular class, as laboring men at moderate work, and find to how many average men of this class the people nourished by each dietary would be equivalent in their demands for nutrients, we should simply have to divide the total quantity of nutrients supplied per day by this equivalent number of men to get the quantities per man per day. The results thus obtained for the several dietaries would, when compared with each other and with accepted standards, give us what we seek.

We are of the opinion that the experimental data on record in European work, if rightly collated and worked up, would give a basis for at least an approximate estimate of the comparative requirements of the several classes of persons into which those nourished by the food of these dietaries would most properly be divided. Indeed, the current standards for daily dietaries will help in arriving at such a basis. Thus, the standards of Voit and the Munich school of physiologists call for proportions of nutrients, with estimated potential energy, as follows:

Persons.	Protein.	Fats.	Carbo-hydrates.	Potential energy.
	Grams.	*Grams.*	*Grams.*	*Calories.*
Children to 1½ years old	28	37	75	767
Children, 6 to 15 years old	75	43	325	2,041
Woman at ordinary work	92	44	400	2,426
Laboring man at moderate work	118	56	500	3,055

We may take the relative quantities of potential energy as the basis of our calculations. The figures are in about the following relative proportions. We interpolate an assumed value for children from 6 to 2 years of age.

Estimated relative quantities of potential energy in nutrients required by persons of different classes.

Laboring man at moderate work	10
Woman at ordinary work	8
Child, 15 to 6 years old	7
Child, 6 to 2 years old	5
Child, under 2 years old	2½

The application of these figures is simple. The food of dietary A, 1, for instance, suffices for 77 persons (factory operatives), 66 males and 11 females. The figures allot to 1 working woman 0.8 as much nutritive material as to 1 laboring man at moderate work. This would make the 11 women equivalent to (8.8) 9 men, which added to 66 would make the whole 77 persons equal to 75 men. The 77 persons during 30 days, the time covered by the dietary, would be equal in requirements to 1 man for 2,250 days. The estimates in the dietaries hereinafter presented are made in this way.

The details of one of the dietaries are quoted to show the nature of the statistics collated by the bureau, and the way in which the quantities of nutrients were estimated. For statements of the ways in which the quantities were estimated from the costs in the cases in which they were not given, and for other details, the reader is referred to the original report, where the details of each of the 30 dietaries are given in full.

TABLE 17.—*Dietary. Series A, No. 1.*

[*Description.*—Boarding-house in Lowell, Mass., of 77 persons, 66 males and 11 females. Boarders, mill operatives. Time, 1 month. Estimated as equivalent in demands for nutrients to 75 laboring men at moderate work for 30 days, or 1 man for 2,250 days.]

FOOD MATERIALS AND NUTRIENTS.

Food materials.				Nutrients.		
Kinds.	Prices per pound.	Quantities.	Costs.	Protein.	Fats.	Carbohydrates.
	Cents.	*Pounds.*		*Pounds.*	*Pounds.*	*Pounds.*
Beef:						
Roast	10	400	$40.00	60.4	79.9	....
Steak	14	272	38.08	39.4	42.4	....
Corned	7	350	24.50	40.3	99.8	....
Tongue	10	62	6.20	9.2	9.5	....
Stew	5	167	8.35	23.4	52.3	....
Tripe	6	20	1.20	4.2	.2	....
Pork, roast	10	150	15.00	17.1	54.3	....
Ham	11	160	17.60	23.4	54.9	....
Salt pork	10	70	7.00	2	53.6	....
Lard	8	260	20.80	....	257.4	....
Haddock	7	168	11.76	13.9	.2	....
Halibut	12	50	6.00	7.6	2.1	....
Mackerel	3	40	1.20	4	1.6	....
Salt fish (cod)	4½	50	2.25	8	.2	....
Total meats, fish, etc	....	2,219	199.94	252.9	708.4	....
Milk	2	3,024	60.48	102.8	111.9	145.2
Cheese	11	63.5	6.98	17.2	22.5	1.5
Butter	22 and 10	291	54.54	2.9	254.6	1.5
Eggs	14	107	14.82	12.4	10.9	.6
Total dairy products and eggs	....	3,485.5	136.82	135.3	399.9	148.8
Flour	3	1,568	47.04	174	17.2	1,182.3
Sugar	7½	600	45.00	....	....	580.2
Molasses	4½	99	4.50	....	....	70.3
Beans	3	124	3.74	28.8	2.6	66.6
Rice	8	25	2.00	1.9	.1	19.9
Oatmeal	4	25	1.00	3.8	1.8	16.8
Potatoes	1	2,520	25.20	47.9	5.0	463.7
Squash	1½	250	3.75	1.3	.3	13.3
Onions	2	26	50	0.3	....	2
Beets	5.9	90	50	1.6	.1	9
Turnips	5.6	120	1.00	1.1	.2	6.1
Tomatoes	5.6	120	1.00	1.6	.4	5.4
Apples	1⅔	300	5.00	.9	....	32.7
Raisins	12½	24	3.00	.6	.1	15.1
Currants	10	15	1.50	.3	....	9.5
Corn starch	9	12	1.08	....	....	11
Crackers	5	48	2.40	5.1	4.8	34
Total vegetable food	....	5,966	148.21	269.2	32.6	2,537.9
Total animal food	....	5,704.5	336.76	388.2	1,108.3	148.8
Total food	....	11,670.5	484.97	657.4	1,140.9	2,686.7

TABLE 17.—*Dietary. Series A, No. 1*—Continued.

FOOD MATERIALS AND NUTRIENTS—Continued.

Food materials.				Nutrients.		
Kinds.	Prices per pound.	Quantities.	Costs.	Protein.	Fats.	Carbohydrates.
	Cents.	*Pounds.*		*Pounds.*	*Pounds.*	*Pounds.*
Meats, fish, etc., per man per day	....	.09	$0.09	.11	.31	....
Dairy products and eggs, per man per day	....	1.55	.06	.06	.18	.07
Animal food, per man per day	....	2.54	.15	.17	.49	.07
Vegetable food, per man per day	....	2.65	.07	.12	.01	1.13
Total food, per man per day	....	5.19	.22	.29	.50	1.20

Results.—The three tables which follow recapitulate the details of fifteen of the dietaries.

Tables 18 and 19 recapitulate the statistics of the dietaries, as explained by their titles.

Table 20 summarizes in shorter form the principal results set forth in the two preceding.

TABLE 18.—*Recapitulation of statistics of dietaries.*

PERSONS, EMPLOYMENTS, WAGES, ETC., AND QUANTITIES AND COSTS OF FOOD.

Series and No. of dietary.	Persons, employments, wages, etc.				Food materials per man per day.										Number of dietary.
					Quantities.					Costs.					
			Wages per day.		Animal food.					Animal food.					
	Dietaries.	Occupation.	Males.	Females.	Meats, fish, etc.	Dairy products and eggs.	Total.	Vegetable food.	Total food.	Meats, fish, etc.	Dairy products and eggs.	Total.	Vegetable food.	Total food.	
	MISCELLANEOUS, MASSACHUSETTS.				*Lbs.*	*Lbs.*	*Lbs.*	*Lbs.*	*Lbs.*	*Cts.*	*Cts.*	*Cts.*	*Cts.*	*Cts.*	
A 11	Family, East Cambridge	Father, glass blower	[1]$4.00		0.66	0.82	1.48	2.97	4.45	9	7	16	9	25	A 11
	Boarding house:														
A 1	Lowell	Mill operatives			.99	1.55	2.54	2.65	5.19	9	6	15	7	22	A 1
A 7	Lynn	Operatives, etc[2]			.71	.91	1.62	3.48	5.10	10	5	15	9	24	A 7
A 2	Lowell	Mill operatives			.98	1.29	2.27	2.66	4.93	10	5	15	7	22	A 2
A 9	Family, Boston	Husband, machinist	[3]3.52		1.36	1.64	3	4.17	7.17	24	12	36	11	47	A 9
	Average of 7 dietaries				.88	1.29	2.17	3.02	5.19	11	6	17	8	25	
	FRENCH CANADIAN, MASSACHUSETTS.														
B 6	Family, Holyoke	Mill operatives	1.35	$0.90	.46	.21	.67	2.36	3.02	7	3	10	7	17	B 6
B 4	Boarding house, Holyoke[4]	Mill operatives	1.25	90	.95	.80	1.75	2.76	4.51	12	6	18	6	24	B 4
B 1	Family, Lawrence	Mill operatives	(—[5])	90	.92	.66	1.58	3.01	4.50	14	4	18	9	27	B 1
B 5	Boarding house, Holyoke	Mill operatives	(—[6])	(7)	1.65	.39	1.44	4.25	5.69	14	5	19	9	28	B 5
B 10	Family, Lowell	Men, blacksmiths[8]	2.00[9]	1.00	1.28	1.51	2.79	4.40	7.19	18	11	29	10	39	B 10
	Average of 10 dietaries				.81	.70	1.51	3.44	4.95	11	5	16	8	24	
	FRENCH CANADIAN, CANADA.														
C 18	Boarding house, Rivière du Loup	All laboring people			.36	.19	.55	1.65	2.20	4	2	6	5	11	C 18
	Family:														
C 12	St. John	All laboring people			.35	.44	.79	2.02	2.81	3	3	6	5	11	C 12
C 26	Richmond	All laboring people			.63	.45	1.08	2.82	3.90	5	3	8	7	15	C 26
C 24	Sherbrooke	All laboring people			1.13	.55	1.68	2.43	4.11	9	4	13	6	19	C 24
C 6	Quebec	All laboring people			.57	.98	1.55	2.25	3.80	5	5	10	6	16	C 6
	Average of 13 dietaries				.52	.45	.97	2.49	3.46	5	3	8	6	14	

[1] Wages $24 per week, work exhausting. [2] Operatives in shoe factory, dressmakers, clerks. [3] $19.50 per week. [4] Board per week, $2.75 for men, $2 for women. [5] $1.25 to $1.75 per day. [6] $1.25 to $1.50 per day. [7] 90 cents to $1 per day. [8] Woman mill operative. [9] $600 per year.

TABLE 18.—*Recapitulation of statistics of dietaries*—Continued.

QUANTITIES OF NUTRIENTS ESTIMATED PER MAN PER DAY.

Series and No. of dietary.	Dietaries.	Nutrients supplied by different classes of food materials.										Total nutrients supplied.						Of every 100 parts of protein the different food materials furnish parts as below.			
		Protein.				Fats.				Carbohydrates.		Hundredths of a pound.			Grams.						
		Animal food.				Animal food.												Animal food.			
		Meats, fish, etc.	Dairy products and eggs.	Total.	Vegetable food.	Meats, fish, etc.	Dairy products and eggs.	Total.	Vegetable food.	Dairy products and eggs.	Vegetable food.	Protein.	Fats.	Carbohydrates.	Protein.	Fats.	Carbohydrates.	Meats, fish, etc.	Dairy products and eggs.	Total.	Vegetable food.
	MISCELLANEOUS, MASSACHUSETTS.	*Lbs.*	*Lbs.*	*Lbs.*	*Lbs.*	*Lbs.*	*Lbs.*	*Lbs.*	*Lbs.*	*Lbs.*	*Lbs.*	*Lbs.*	*Lbs.*	*Lbs.*	*Grms.*	*Grms.*	*Grms.*	*P. ct.*	*P. ct.*	*P. ct.*	*P. ct.*
A 11	Family	0.07	0.03	0.10	0.11	0.13	0.14	0.27	0.02	0.03	1.03	0.21	0.29	1.06	95	132	481	33	14	47	53
A 1	Boarding house	.11	.06	.17	.12	.31	.18	.49	.01	.07	1.13	.29	.50	1.20	132	227	545	38	21	59	41
A 7	.....do	.09	.04	.13	.12	.19	.12	.31	.02	.04	1.11	.25	.33	1.15	114	150	522	36	16	52	48
A 2	.....do	.11	.05	.16	.13	.30	.13	.43	.01	.06	1.15	.29	.44	1.21	132	200	549	38	17	55	45
A 9	Family	.17	.08	.25	.15	.31	.21	.52	.04	.07	1.29	.40	.56	1.36	182	254	617	42	20	62	38
	Average of 7	.11	.05	.16	.12	.24	.15	.39	.02	.06	1.11	.28	.41	1.17	127	186	531	39	18	57	43
	FRENCH CANADIANS, MASSACHUSETTS.																				
B 6	Family	.04	.02	.06	.12	.21	.06	.27	.02	......	1.09	.18	.29	1.09	82	132	495	22	11	33	67
B 4	Boarding house	.09	.04	.13	.08	.49	.09	.58	.01	.03	.69	.21	.59	.72	95	268	327	43	19	62	38
B 1	Family	.10	.03	.13	.12	.33	.07	.40	.01	.03	1.12	.25	.41	1.15	114	186	522	40	12	52	48
B 5	Boarding house	.10	.02	.12	.19	.47	.10	.57	.02	.01	1.21	.31	.59	1.22	141	268	554	32	7	39	61
B 10	Family	.15	.08	.23	.21	.48	.15	.63	.04	.05	1.70	.44	.67	1.75	200	304	795	34	18	52	48
	Average of 10	.08	.04	.12	.14	.34	.09	.43	.02	.02	1.19	.26	.45	1.21	118	204	549	31	15	46	54
	FRENCH CANADIANS, CANADA.																				
C 18	Boarding house	.04	.01	.05	.11	.09	.08	.17	.02	......	.85	.16	.19	.85	73	86	386	25	6	31	69
C 12	Family	.05	.01	.06	.12	.07	.11	.18	.02	.02	.93	.18	.20	.95	82	91	431	28	5	33	67
C 26	.....do	.09	.02	.11	.12	.10	.10	.20	.02	.02	1.13	.23	.22	1.15	104	100	522	39	9	48	52
C 24	.....do	.14	.02	.16	.17	.17	.17	.34	.03	.02	1.16	.33	.37	1.18	150	168	536	42	6	48	52
C 6	.....do	.08	.04	.12	.17	.08	.12	.20	.02	.04	1.09	.29	.22	1.13	132	100	513	27	14	41	59
	Average of 13	.07	.02	.09	.15	.10	.11	.21	.03	.02	1.14	.24	.24	1.16	109	109	527	29	8	37	63

TABLE 19.—*Persons stated to be nourished by food of dietaries, and estimated numbers of laboring men at moderate work who would require the same quantities of nutrients.*

Series and No. of dietary.	Persons reported.	Classification.					Total number of persons.	Estimated equivalent to laboring men.
		Adults.		Children.				
		Males.	Females.	15 to 6 years.	6 to 2 years.	Under 2 years.		
	MISCELLANEOUS, MASSACHUSETTS.							
A 11	Father, mother, 1 other adult female, and 3 children of 5, 11, and 12 years..	1	2	2	1	....	6	4 1/3
	Boarding house:							
A 1	66 males and 11 females............	66	11	....	....	....	77	75
A 7	20 males and 16 females............	20	16	....	....	....	36	33
A 2	10 males and 60 females............	10	60	....	....	....	70	58
A 9	Husband and wife.....................	1	1	....	....	....	2	1 4/5
	FRENCH CANADIAN, MASSACHUSETTS.							
B 0	Father, mother, 2 adult children.[1] and 2 children of 9 and 12½ years.........	2	2	2	....	....	6	5
B 4	Boarding house, 8 men, 7 women, and 3 children	8	7	3	....	....	18	15 1/3
B 1	Father, mother, and 4 adult children, 1 female	4	2	....	....	....	6	5 3/5
B 5	Boarding house, 6 males and 4 females, ages 16 to 40 years	6	4	....	....	....	10	9 1/2
B 10	Two brothers[2] and a sister, adults.....	2	1	....	....	....	3	3
	FRENCH CANADIAN, CANADA.							
C 18	Boarding house, 15 adults	8	7	....	....	....	15	13 1/2
C 12	Father, mother, and 8 children, 2 to 13 years old	1	1	5	3	....	10	6 3/5
C 26	Father, mother, and 3 children of 9, 12, and 14 years........................	1	1	3	....	....	5	4
C 24	Father, mother, and 2 children, 6 months and 5 years old	1	1	....	1	1	4	2 1/2
C 6	Father, mother, and 6 children, 1 to 12 years old	1	1	3	2	1	8	5

[1] One male and one female.
[2] The men, blacksmiths, were at rather severe work, hence the 2 with 1 woman are estimated as equivalent to 3 men at moderate manual labor.

TABLE 20.—*Summary of statistics of dietaries.*[1]

QUANTITIES OF FOOD MATERIALS.

Food materials.	Series A, miscellaneous, Massachusetts.			Series B, French Canadian, Massachusetts.			Series B, French Canadian, Canada.		
	Maximum.	Minimum.	Average.	Maximum.	Minimum.	Average.	Maximum.	Minimum.	Average.
	Lbs.	*Lbs.*	*Lbs.*	*Lbs.*	*Lbs.*	*Lbs.*	*Lbs.*	*Lbs.*	*Lbs.*
Meats, fish, etc........................	1.36	0.63	0.88	1.28	0.46	0.81	1.13	0.35	0.52
Milk, butter, cheese, and eggs	1.70	.82	1.29	1.51	.21	.70	.98	.14	.45
Total animal food	3.00	1.48	2.17	2.79	.67	1.51	1.68	.54	.97
Vegetable food......................	4.17	2.38	3.02	5.65	2.35	3.44	3.65	1.65	2.49
Total food....................	7.17	4.12	5.19	7.26	3.02	4.95	4.89	2.20	3.46

[1] The figures for maximum and minimum indicate the largest and smallest quantities in any single dietary, and those for average, the averages of all the dietaries of each series. Thus, the largest quantities of meat, etc., per man per day in any of the dietaries of Series A was 1.36 pounds, the smallest 0.63 pounds, and the average of the 7 dietaries of this series examined was 0.88 pound. The largest amount of total food in any single dietary of this series was 7.17 pounds, the smallest 4.12 pounds, and the average 5.19 pounds. That the figures for total do not always equal the corresponding sum (for instance, the total animal food, maximum, Series A, is less than the sum of the figures for meats, fish, etc., and for milk, butter, cheese, and eggs) is due to the fact that the factors which would make up the sum are from different dietaries, while the figures for total are the maximum, minimum, etc., for individual dietaries.

TABLE 20.—*Summary of statistics of dietaries*[1]—Continued.

COSTS OF FOOD MATERIALS.

Food materials.	Series A, miscellaneous, Massachusetts.			Series B, French Canadian, Massachusetts.			Series B, French Canadian, Canada.		
	Maximum.	Minimum.	Average.	Maximum.	Minimum.	Average.	Maximum.	Minimum.	Average.
	Cts.	*Cts.*	*Cts.*	*Cts.*	*Cts.*	*Cts.*	*Cts.*	*Cts.*	*Cts.*
Meats, fish, etc.	24	6	11	18	6	11	9	3	5
Milk, butter, cheese, and eggs	12	4	6	11	3	5	7	1	3
Total animal food	36	10	17	29	10	16	13	5	8
Vegetable food	11	6	8	13	6	8	6	5	6
Total cost	47	16	25	39	17	24	19	11	14

NUTRIENTS IN FOOD MATERIALS.

Food materials.	Series A, Maximum.	Series A, Minimum.	Series A, Average.	Series B (Mass.), Maximum.	Series B (Mass.), Minimum.	Series B (Mass.), Average.	Series B (Canada), Maximum.	Series B (Canada), Minimum.	Series B (Canada), Average.
	Lbs.	*Lbs.*	*Lbs.*	*Lbs.*	*Lbs.*	*Lbs.*	*Lbs.*	*Lbs.*	*Lbs.*
Protein	0.40	0.21	0.28	0.44	0.18	0.26	0.33	0.16	0.24
Fats	.56	.29	.41	.67	.28	.45	.39	.16	.24
Carbohydrates	1.36	1.05	1.17	1.75	.72	1.21	1.59	.85	1.16
Total nutrients	2.32	1.56	1.86	2.86	1.52	1.92	2.29	1.20	1.64
Percentages of animal protein in total protein food	64	47	57	62	33	46	48	29	37

[1] The figures for maximum and minimum indicate the largest and smallest quantities in any single dietary, and those for average, the averages of all the dietaries of each series. Thus, the largest quantities of meats, etc., per man per day in any of the dietaries of Series A was 1.36 pounds, the smallest 0.63 pound, and the average of the 7 dietaries of this series examined was 0.88 pound. The largest amount of total food in any single dietary of this series was 7.17 pounds, the smallest 4.12 pounds, and the average 5.19 pounds. That the figures for total do not always equal the corresponding sum (for instance, the total animal food, maximum, Series A, is less than the sum of the figures for meats, fish, etc., and for milk, butter, cheese, and eggs) is due to the fact that the factors which would make up the sum are from different dietaries, while the figures for total are the maximum, minimum, etc., for individual dietaries.

Discussion of results.—The discussion of the dietary statistics involves details of local interest, but the following quotations from the report will serve to show some of the ways in which such information can be made useful. It will be remembered that Series A, miscellaneous, Massachusetts, includes dietaries of factory and mill operatives, mechanics, and a few clerks, dressmakers, etc., of various nationalities, in Lowell, Lawrence, Lynn, East Cambridge, and Boston. Series B, French Canadians, Massachusetts, includes factory operatives and a few mechanics of Canadian origin, working in Massachusetts. Series C, French Canadians, Canada, includes similar people, mainly or entirely laboring classes in Canada. In other words, the people had only very moderate incomes, and the majority were factory operatives, a class of whom we ordinarily think as living on rather a low plane of material comfort. The following statements are from the report referred to:

In the following table (21) the averages of the analyses of dietaries are succinctly set forth:

TABLE 21.—*Averages of statistics of dietaries.*

QUANTITIES OF FOOD MATERIALS.

Food materials.	Series A, miscellaneous, Massachusetts.	French Canadian.	
		Series B, Massachusetts.	Series C, Canada.
	Pounds.	*Pounds.*	*Pounds.*
Animal	2.17	1.51	0.97
Vegetable	3.02	3.44	2.49
Total	5.19	4.95	346

COSTS OF FOOD MATERIALS.

	Cents.	*Cents.*	*Cents.*
Animal	17	16	8
Vegetable	8	8	6
Total	25	24	14

NUTRIENTS IN FOOD MATERIALS.

	Grams.	*Grams.*	*Grams.*
Protein	127	118	109
Fats	186	204	109
Carbohydrates	531	549	527
Total	844	871	745
Animal protein in 100 of total protein	57	46	37

DIETARY OF A BOARDING HOUSE IN CONNECTICUT.

The Storrs (Conn.) Experiment Station has lately cooperated with the United States Department of Labor in the study of a number of dietaries. Some of the results of the studies have been published by the station in its reports for 1891–1893. The work has been done under the writer's direction in the chemical laboratory of Wesleyan University. The following account of the study of the dietary of a boarding house was published in the report of the station for 1891.

The details of the investigation were carefully carried out by Mr. H. B. Gibson, at that time assistant chemist to the station. During the whole period of the investigation Mr. Gibson boarded at the house and kept account of everything purchased and of kitchen and table wastes.

The general plan of the investigation included account of all food materials of nutritive value in the house at the beginning, that purchased during and that which remained at the end of the experiment. In addition to this all the kitchen and table wastes of the food were collected, taken to the laboratory, and there weighed and analyzed. The amount of different food materials on hand at the beginning and received during the experiment were added; from this sum the amounts remaining at the end were subtracted. This gave the amount of each material actually used. From the amounts thus obtained and the composition of each material as shown by analysis, the amounts of the nutritive ingredients were estimated. From this were subtracted the amounts of nutrients in the waste, and thus the amounts of nutrients in the food actually eaten were learned.

Duration of the experiment.—The dietary commenced with supper, October 20, 1890, and continued until after dinner of November 19, a period of 30 days.

Members of the family and meals eaten.—During most of the time the family consisted of 13 men and 7 women. It very rarely happened that all of the family took all three meals at the house any given day. There were occasional visitors, and in this way once or twice the total number of meals taken per day was larger than the family alone would have required. The sex, approximate age, and occupation of each member of the family, as it was constituted most of the time, were as follows:

Men:	
Machinists, 30 to 40 years of age	4
Machinist, about 35, after October 25	1
Harness maker, about 70	1
Hired men, one old, the other middle aged	2
Proprietor of the house, about 70	1
Manufacturers, one about 60, the other about 30	2
Chemist, about 27	1
Reporter for newspaper, about 20, after October 27	1
Total	13
Women:	
Housekeeper, about 30	1
Cook, about 45	1
Table girl, about 20	1
Doing no manual labor, about 30, 55, 55, and 70	4
Young lady at house 4 days, doing no labor	1
Total	8

Of the 13 men, 3 were counted as "hearty eaters," and 6 more as having decidedly good appetites.

An account of the number of meals taken each day was kept, and is given in the following table:

Number of meals taken at the house each day.

Date.	Men.	Women.	Date.	Men.	Women.	Date.	Men.	Women.
Oct. 20	10	7	Oct. 31	37	20	Nov. 11	38	13
21	32	21	Nov. 1	42	18	12	38	16
22	29	23	2	40	16	13	38	14
23	32	25	3	40	16	14	37	16
24	32	23	4	36	16	15	38	16
25	30	17	5	39	18	16	37	18
26	25	21	6	38	19	17	37	18
27	34	19	7	36	18	18	37	19
28	35	19	8	37	16	19	21	11
29	35	18	9	33	14			
30	34	18	10	33	13	Total	1,060	536

The actual number of meals taken at the house during the 30 days of the experiment was 1,596, of which 1,060 were eaten by men and 536 by women. Assuming, in accordance with the dietary standards given in the succeeding chapter, that on the average one woman ate eight-tenths as much as one man, this would reduce the whole number of meals to an equivalent of 1,489 for the 30 days, or 3 meals per day for 496 days for 1 man.

Food used during the experiment.—The actual amount of food and of nutrients in the food used during the dietary is shown in Table 22, which follows. The weights of the food materials are as they were purchased and used; that is, they include bone and other refuse except where specified.

The first three columns in the table contain the percentages of protein, fat, and carbohydrates used in computing the amounts of these nutrients in the different food materials.

In nearly all cases analyses were made of specimens of the food materials actually used in the dietary, or of specimens so nearly identical with those used that they might properly serve for the analysis. The analysis was omitted in a few cases, but only where previous analyses were thought to furnish reasonably accurate data for estimates of the composition.

TABLE 22.—*Food materials used in dietary of boarding house for 13 men and 8 women during 30 days.*

Food materials.	Percentage composition.			Weights used.			
	Protein.	Fat.	Carbohydrates.	Total food materials.	Nutrients.		
					Protein.	Fat.	Carbohydrates.
ANIMAL FOODS.							
Beef:	*Per cent.*	*Per cent*	*Per cent.*	*Grams.*	*Grams.*	*Grams.*	*Grams.*
Short steak	18.6	19.2	....	5,780	1,075	1,109	....
From soup bones	16.8	30.3	....	1,100	185	333	....
Chuck ribs	13.2	19.7	....	18,270	2,412	3,599	....
Do	15.2	28.7	....	3,290	500	944	....
Do	16.3	13.3	....	2,720	443	362	....
Do	15.6	17.8	....	2,780	434	495	....
Shoulder	17.3	8.8	....	2,610	452	230	....
Roast, free from bone	17.1	19.9	....	2,780	475	552	....
Do	16.1	32.9	....	2,500	403	823	....
Do	16	32	....	2,160	346	691	....
Do	16.8	30.3	....	2,670	449	809	....
Do	15.2	36.6	....	2,720	413	996	....
Suet	.9	94.6	....	340	3	322	....
Corned	13.2	21.3	....	14,050	1,855	2,993	....
Do	11.7	37.2	....	9,370	1,096	3,485	....
Corned, canned	24.5	20.2	....	11,168	2,736	2,256	....
Dried	29.9	4.5	....	3,400	1,017	1,153	....
Tripe	13.5	1.9	....	2,950	398	56	....
Bologna sausage	18.8	15.8	....	3,970	746	627	....
Total	....	....	....	94,628	15,438	21,835	....
Veal:							
Shoulder	18.3	5.5	....	9,750	1,784	536	....
Miscellaneous cuts	14.9	10.2	....	3,400	507	347	....
Total	....	....	....	13,150	2,291	883	....
Pork:							
Ribs	13.5	25.4	....	3,400	459	864	....
Ribs and shoulder	14.1	25.6	....	6,410	904	1,641	....
Shoulder	13.8	28.9	....	1,470	203	425	....

TABLE 22.—*Food materials used in dietary of boarding house for 13 men and 8 women during 30 days*—Continued.

Food materials.	Percentage composition.			Weights used.			
					Nutrients.		
	Protein.	Fat.	Carbohydrates.	Total food materials.	Protein.	Fat.	Carbohydrates.
ANIMAL FOODS—continued.							
Pork—Continued.	*Per cent.*	*Per cent*	*Per cent.*	*Grams.*	*Grams.*	*Grams.*	*Grams.*
Chops	13.2	31.9	....	8,560	1,130	2,731	....
Salt, fat	.9	82.8	....	1,250	113	1,035	....
Ham, smoked	14.8	34.6	....	22,340	3,306	7,730	....
Sausage	11.2	45.4	....	8,280	928	3,759	....
Lard	....	90	....	6,350	....	5,715	....
Total	....	....	....	58,060	7,043	23,900	....
Lamb:							
Ribs	16.7	22.1	....	4,650	777	1,028	....
Shoulder	13.5	22.8	....	11,790	1,592	4,280	....
Leg	15.6	12.6	....	6,340	989	789	....
Total	....	....	....	22,780	3,358	6,097	....
Fowl and chicken	15.1	1.2	....	14,290	2,158	171	....
Fish:							
Flounder, dressed	6.3	.3	....	4,200	265	13	....
Haddock, dressed	9.9	.2	....	6,860	679	14	....
Cod, salt	16	.4	....	790	126	3	....
Round clams, shell contents	6.5	.4	....	3,910	254	16	....
Total	....	....	....	15,760	1,324	46	....
Dairy products, etc.:							
Butter	....	80	....	34,190	....	27,352	....
Milk	3.5	4.5	4.9	191,400	6,699	8,613	9,379
Cheese	27.1	35.5	2.3	1,130	306	401	26
Total	....	....	....	226,720	7,005	36,366	9,405
Eggs	12.2	10.2	....	12,560	1,532	1,281	....
Total animal foods	....	....	....	457,948	40,149	90,579	9,405
VEGETABLE FOODS.							
Onions (10 per cent refuse)	1.4	.3	10.1	6,120	86	18	618
Sweet potatoes (12.5 per cent refuse)	1.5	.4	26	30,564	458	122	7,947
Potatoes (15 per cent refuse)	2.1	.1	17.9	98,175	2,062	98	17,573
Squash (50 per cent refuse)	.9	.2	10.1	3,400	31	7	343
Cabbage (15.5 per cent refuse)	1.5	.2	5.7	14,895	223	30	849
Turnips (30 per cent refuse)	1.2	.2	8.2	38,150	458	76	3,128
Canned corn	3	1.3	22.5	3,870	116	50	871
Canned peas	4.3	.3	11.4	3,945	170	12	449
Canned tomatoes	1	.2	3.7	5,190	52	10	192
Dried beans	22.2	1.4	60.3	4,080	906	57	2,460
Apples (25 per cent refuse)	.3	.4	15.9	33,875	102	136	5,385
Cranberries	.4	.9	10.9	2,640	11	24	288
Grapes (25 per cent refuse)	1.7	1.7	21.3	10,120	172	172	2,156
Flour	13.5	1.3	74	124,500	16,908	1,619	92,130
Buckwheat flour	6.9	1.4	76.1	1,810	125	25	1,377
Corn starch, tapioca, and pearl barley	....	....	97.5	1,130	....	....	1,102
Hominy	8.2	.4	77.4	2,270	186	9	1,757
Oatmeal	15.1	7.1	68.2	1,470	222	104	1,002
Crackers	10.7	9.9	68.8	1,360	146	135	936
Sugar (water-free)	....	....	100	60,780	....	....	60,780
Molasses and sirup (63 per cent sugar)	....	....	63	6,630	....	....	418
Total	....	....	....	454,974	22,434	2,704	201,761
Total animal and vegetable foods	....	....	....	912,922	62,583	92,262	211,166

NOTE.—100 grams = 3.5 ounces, or 0.22 pounds. 1 ounce = 28.35 grams. 1 pound = 453.6 grams.

The data of Table 22 are summarized in Table 23, in which the weights are stated in pounds as well as in grams. As before explained, it was estimated that the food consumption was equivalent to that for one man for a period of 496 days. The table gives the quantities estimated per man per day on this basis. These figures apply to the food as purchased.

TABLE 23.—*Weights of food materials and of nutritive ingredients used.*

RECAPITULATION OF DIETARY OF BOARDING HOUSE.

Food materials.	Food materials.	Nutrients.			Food materials.	Nutrients.		
		Protein.	Fats.	Carbohydrates.		Protein.	Fats.	Carbohydrates.
FOR 21 PERSONS, 30 DAYS.	*Grams.*	*Grams.*	*Grams.*	*Grams.*	*Lbs.*	*Lbs.*	*Lbs.*	*Lbs.*
Meats, etc	202,908	30,288	52,880	....	446	67	116	....
Fish	15,760	1,324	46	....	35	3	....	....
Dairy products:								
Butter	34,190	....	27,352	....	75	....	60	....
Cheese	1,130	306	401	26	3	1	1	....
Milk	191,400	6,699	8,613	9,379	421	15	19	21
Total	226,720	7,005	36,366	9,405	499	16	80	21
Eggs	12,560	1,532	1,281	....	28	3	3	....
Total animal food	457,948	40,149	90,579	9,405	1,008	89	199	21
Vegetable food	454,974	22,434	2,704	201,761	1,001	49	6	444
Total food	912,922	62,583	93,283	211,166	2,009	138	205	465
PER MAN PER DAY.								
Meats, etc	409	61.1	106.6	....	.9	....	....	....
Fish	32	2.7	.1	....	.1	....	....	....
Dairy products:								
Butter	69	....	55.2	....	.2	....	....	....
Cheese	2	.6	.8	....	....	....	....	....
Milk	386	13.5	17.4	18.9	.8	....	....	....
Total	457	14.1	73.4	18.9	.1	....	....	....
Eggs	25	3.1	2.6	....	.1	....	....	....
Total animal food	923	81	182.7	18.9	2.1	.18	.40	.04
Vegetable food	917	45.2	5.5	406.9	2	.10	.01	.90
Total food	1,840	126.2	188.2	425.8	4.1	.28	.41	.94

Table 23 tells its story so plainly as to require little comment. It will, however, be interesting to note the proportion of the several kinds of food materials in the dietary and the relative proportions of the several nutrients furnished by each. These are expressed in percentages in Table 24. Thus in the column under "Total food materials," it will be observed that the meats make up 22.2 per cent, the milk 21 per cent, and the butter and cheese sufficient to make the total dairy products 24.8 per cent; the animal food altogether 50.1, and the vegetable food 49.9 per cent of the total food purchased. Of the whole protein 48.4 per cent was supplied in the meats, 2.2 per cent in the fish, and so on.

TABLE 24.—*Percentages of food materials of different classes and of nutrients furnished by each class.*

Food materials.	Total food materials.	Nutrients.		
		Protein.	Fats.	Carbohydrates.
	Per cent.	*Per cent.*	*Per cent.*	*Per cent.*
Meats, etc	22.2	48.4	56.7	....
Fish	1.7	2.2	.1	....
Dairy products:				
Butter	3.7	....	29.3	....
Cheese	.1	.5	.4	.1
Milk	21	10.7	9.2	4.4
Total	24.8	11.2	38.9	4.5
Eggs	1.4	2.4	9.2	....
Total animal food	50.1	64.2	97.1	4.5
Vegetable food	49.9	35.8	2.9	95.5
Total food	100	100	100	100

Table and kitchen refuse and waste.[1]—These were made up of bones of meat, parings of potatoes, etc., left in preparing the food for the table, and the scraps of food left on the table uneaten. The percentages of refuse of vegetables and meats were obtained from experiment as described in the footnote. All of the table and kitchen refuse and waste, except parings of vegetables and bones of meat, were carefully collected after each meal and taken to the laboratory. It was there freed from bone and other inedible materials, and then dried for 24 hours or more in large evaporating dishes in a water oven at 98° C., as usual in preparing for analysis. While still hot, the fat was removed as far as possible, so as to leave the samples in better condition for

[1] The terms "refuse" and "waste" and the ways of determining the amounts demand a word of explanation. The weights given for the food materials in Table 22 are those found by weighing them as received, and with few exceptions include both edible portion and refuse. The exceptions are stated in the table, e. g., "Beef, roast, free from bone." The proportions of refuse in the meats were found by weighing the refuse and edible portion of the specimens analyzed. The percentages of refuse and edible portion thus found are given in the tables of composition of animal foods, in the chapter on "Composition of food materials" of the present report. Those figures for refuse accordingly represent actually inedible material and not waste in the proper sense, i. e., valuable material wasted, except in so far as the bone, etc., may be utilized for soup. The figures in Table 22 for refuse of the vegetable foods, e. g., parings of potatoes and turnips, stem and outside leaves of cabbage, rind and seeds, etc., of squash, were determined approximately by weighings at the house. They represent not only the actually inedible material, but also more or less of edible material removed with it. For instance, it was estimated from weighings of the parings of the potatoes that out of 100 pounds, 15 pounds were removed in the kitchen and only 85 pounds served on the table. These 15 pounds were made up in part of the skin and adhering earth, which might be properly called refuse. The rest was edible material, which, if it had been saved and served, would have been as wholesome as that which went on the table, and may be appropriately designated as waste. The terms refuse and waste are here used somewhat indiscriminately. Perhaps a correct terminology would make the refuse consist of the actually inedible material and waste, the latter term covering the edible material rejected with the refuse. While the use of terms is unimportant, the subject itself is an important one for American household economy.

grinding. The water-free fat thus removed amounted in all to 2,639 grams. It is accounted for in Table 25. The partly dried refuse was allowed to accumulate till 5 kilograms or more had accumulated, and was then sampled and prepared for analysis. Eight such lots of material were collected and analyzed. The weights of the water-free refuse and its constituents are given in Table 25 herewith.

TABLE 25.—*Percentages and weights of nutrients in table and kitchen wastes.*

Laboratory number.	Percentage composition.				Weight of ingredients.				
					Total water-free substance (nutritive materials).	Nutrients.			
	Protein.	Fats.	Carbohydrates.	Ash.		Protein.	Fats.	Carbohydrates.	Ash.
	Per cent.	*Per cent.*	*Per cent.*	*Per ct.*	*Grams.*	*Grams.*	*Grams.*	*Grams.*	*Grams.*
307	27.44	40.94	28.52	3.10	5,514	1,513	2,258	1,573	170
310	26.61	31.57	37.31	4.31	5,772	1,548	1,822	2,154	248
313	28.56	38.89	28.84	3.71	4,478	1,270	1,742	1,291	166
317	26.50	36.50	32.42	4.58	4,653	1,233	1,698	1,500	213
324	28.38	37.39	30.44	3.79	4,831	1,371	1,806	1,471	183
328	32.75	40.83	21.45	4.97	5,659	1,853	2,311	1,214	281
334	31.31	40.59	23.46	4.64	5,198	1,628	2,110	1,219	241
336	25.94	33.27	36.28	4.51	4,215	1,093	1,402	1,520	191
Fat removed as above explained					2,639	....	2,639	....	....
Total grams					42,959	11,518	17,788	11,960	1,693
Equivalent to pounds					94.4	25.3	39.1	26.3	3.7

NOTE.—100 grams = 3.5 ounces, or 0.22 pounds. 1 ounce = 28.35 grams. 1 pound = 453.6 grams.

TABLE 26.—*Nutrients and potential energy in food purchased, rejected, and eaten.*

	Nutrients.			Potential energy.
	Protein.	Fats.	Carbohydrates.	
WHOLE EXPERIMENT, 21 PERSONS, 30 DAYS.				
Food purchased:	*Grams.*	*Grams.*	*Grams.*	*Calories.*
Animal	40,149	90,579	9,405	1,045,560
Vegetable	22,434	2,704	201,761	944,350
Total	62,583	93,283	211,166	1,989,910
Waste:				
Animal	10,190	17,621	....	205,650
Vegetable	1,328	167	11,960	56,030
Total	11,518	17,788	11,960	261,680
Food actually eaten:				
Animal	29,959	72,958	9,405	839,900
Vegetable	21,106	2,537	189,801	888,310
Total	51,065	75,495	199,206	1,728,210
PER MAN PER DAY.				
Food purchased:				
Animal	81	182.7	18.9	2,110
Vegetable	45.2	5.5	406.9	1,900
Total	126.2	188.2	425.8	4,010
Waste:				
Animal	20.5	35.5	....	410
Vegetable	2.7	.3	24.1	110
Total	23.2	35.8	24.1	520
Food actually eaten:				
Animal	60.4	147.1	18.9	1,690
Vegetable	42.6	5.1	382.8	1,790
Total	103	152.2	401.7	3,480

TABLE 26.—*Nutrients and potential energy in food purchased, etc.*—Continued.

TOTAL FOOD PURCHASED.

	Nutrients.			Potential energy.
	Protein.	Fats.	Carbo-hydrates.	
Food purchased:	*Per cent.*	*Per cent.*	*Per cent.*	*Per cent.*
Animal	64.2	97.1	4.5	52.5
Vegetable	35.8	2.9	95.5	47.5
Total	100	100	100	100
Waste:				
Animal	16.3	18.9	....	10
Vegetable	2.1	.2	5.6	2.7
Total	18.4	19.1	5.6	12.7
Food actually eaten:				
Animal	47.9	78.2	4.5	41.2
Vegetable	33.7	2.7	89.9	43.7
Total	81.6	80.9	94.4	84.9

NOTE.—100 grams = 3.5 ounces, or 0.22 pounds. 1 ounce = 28.35 grams. 1 pound = 453.6 grams.

It will be worth while to estimate, as well as may be, how much of the nutritive material wasted came from the animal and how much from the vegetable food. As there were practically no carbohydrates in any of the animal food materials except milk and cheese, and but little in these, we shall not greatly err in assuming that all the waste carbohydrates came from the vegetable foods. For a rough calculation, it will also be safe to assume that the proportions of protein and fat in the vegetable portion of the waste were the same as in the whole vegetable food. In the latter the weight of the protein is 11.1 per cent and that of fat 1.4 per cent of the weight of the carbohydrates. Taking the same percentages of the weight of the carbohydrates in the total waste as the measure of the protein and fats in the vegetable waste, we have the actual weights of protein and fats in the latter. Subtracting these weights of vegetable protein and fat from the total weights of these ingredients in the waste, we have as the remainder the amounts of animal protein and fat in the whole waste.

The proportions of nutritive ingredients in the food purchased, in the waste, and in the food actually consumed are given in Table 26, on the preceding page. The estimates of animal and vegetable nutrients in the waste are computed as above described, and those of potential energy as explained above.

About one-ninth of the total nutritive ingredients of the food was left in the kitchen and table wastes. The actual waste was worse than this proportion would imply, because it consisted mostly of the protein and fats which are more costly than the carbohydrates. The waste contained nearly one-fifth of the total protein and fat, and only one-twentieth of the total carbohydrates of the food. Or, to put it in another way, the food purchased contained about 23 per cent more protein,

24 per cent more fats, and 6 per cent more carbohydrates than were eaten. And, worst of all for the pecuniary economy or lack of economy, the wasted protein and fats were mostly from the meats which supply them in the costliest form. At the ratio in which the nutrients were actully eaten in this dietary, the protein in the waste would have sufficed 1 man for 112 days; the fats would have supplied him also for 112 days, and the carbohydrates for 30 days.

DIETARIES OF HAND WEAVERS IN ZITTAU, SAXONY.

An investigation of the conditions of living, with reference especially to income and expenditure of a class of poor people, weavers, in Zittau, a district in eastern Saxony, was made in 1885 by Von Schlieben, and reported in a publication[1] of the Saxon Bureau of Statistics. Later, Von Rechenberg, who had been connected for some time with the laboratory of the Agricultural Institute of the University of Leipsic, and had been engaged with Professor Stohmann in the combustion of food materials, became interested to study the food consumption of these people more fully, and has, with the aid of the Royal Saxon Society of Sciences, published the results of his inquiry in considerable detail.[2]

The following summary is taken in part from the report of Von Schlieben, but in the main from that of Von Rechenberg:

Zittau is one of the few localities in Germany where weaving on a large scale is still done by hand. The persons over 16 years of age engaged in this industry, in 1885, numbered 8,000 out of a population of somewhat over 73,000, distributed throughout 65 villages. The industry is an old one; so long ago as 1729 statistics concerning it were gathered by the Government.

The majority of the weavers in 1885 were over 30 years of age and more than half were over 40.

The stuffs woven are linen toweling, bed linen, damask, cotton and woolen cloth, and similar wares.

The hours of work are very protracted, ranging from 5 or 6 a. m. to 8 or 9 p. m. in summer and from 6 or 7 a. m. to 11 p. m. in winter; that is, 13 to 14 hours a day in summer and 14 to 16 hours in winter.

The old-fashioned wooden looms are used. In the weaving of cloth of ordinary width the muscular exertion is not severe, as the work is done while sitting; but some of the cloth is very wide, and in making this severe exertion is required. In fact, the long-continued practice of weaving extra wide cloth with the pressure of the body against the loom often produces serious internal disorders.

From 3 to 5 marks (1 mark equals about 24 cents) per week are earned for weaving the ordinary kinds of cloth, and from 5 to 10 marks for the extra wide and fine wares. The latter sums are, however, the exception.

---

[1] Zeitschrift des k. sächischen statischen Bureaus, 1885, Parts III and IV, pp. 156–190.

[2] Die Ernährung der Handweber in der Amtshauptmannschaft, Leipzig, 1890.

In several cases the reported earnings of a man and wife together aggregated only about 300 marks per year.

The habits and mode of life of these people are very simple. The house often consists of only 3 rooms. The largest is the general living room; in this the weaving is done. The two smaller are used as bedroom and kitchen.

The weaving is done by both the man and woman, in so far as the household duties of the latter permit. The children when not in school help in winding the thread and in other ways.

The extremely weak physical condition of these weavers incapacitates them, as a rule, for farm work; but occasionally the more robust ones work on the farms for a few weeks during harvest time.

The larger part of the earnings of these people is spent for food. In none of the families did the amount paid for food fall below one half of the total income, and in several cases it reached four-fifths. The average expenditure for food was found to be about two-thirds of the total income.

An idea of their condition may be gained from the fact that while the climate is about like that of southern New England, something like $7.50 per year would be the average sum expended per family for fuel for heating, cooking, and other household purposes. Many gather part of the fuel they use in the public forests or on the large estates of families living in this region. Of course they are allowed to collect only such materials as small twigs and branches which have been blown from the trees. Others prepare fuel in summer by kneading together a mixture of cheap brown coal and peat, shaping it into cakes and drying it.

The food of these people is of the simplest kind. Variations from the regular diet are very exceptional. Meat is rarely eaten except on Sunday and then in small quantities, and even among the better class of weavers such luxuries as beer are indulged in only on great occasions.

The diet is not peculiar to this region, but is similar to the so-called "potato diet" so common among German peasants. The term "potato diet" is somewhat misleading. In reality rather more than half of the total food is bread, and only in exceptional cases would the quantity of potatoes be more than one-third of the whole.

The physical condition of these people is far from what might be desired. There is a notable lack of muscular energy. Still the weavers are by no means short lived. When the household duties of the woman are so great as to keep her from the loom, her physical condition is usually superior to that of the man. The young children are more healthy than the older ones.

Notwithstanding the unfavorable conditions of life, the people seem contented with their lot, and Dr. von Rechenberg regards them as much better off than many families of factory operatives in cities like

Leipsic who have much larger incomes, but whose occupations deprive them of home life.

After a general idea of the condition of life of the Zittau band weavers had been gained by discussing the subject with the leading men, employers and employed, and by visiting many families in a number of villages, 52 families were selected as representative of the class which made the best use of the means at their disposal. To these were given blanks to fill out. The character of the questions and answers in these blanks may be inferred from the following translation of one of them. The answers to the questions are in italics:

QUESTION CARD TO AID IN DETERMINING THE CONDITIONS OF A WEAVER'S FAMILY IN DITTELSDORF.

[Family No. 2.]

1. Age of husband, *56*. Age of wife, *71*. Number of children. Age of children (*Childless*).

2. Number and kind of rooms in house? *One living room, 1 bedroom, 1 attic, 1 small cellar.*

3. What are the chief articles of diet—bread or potatoes, or vegetables and flour? *Bread, flour, and potatoes.*

4. How often in the week is meat eaten, and is beef or pork preferred? *Once, pork preferred.*

5. Weekly expenditures for food, etc.?

Expenditures for food, etc., in one week.

Kinds of material purchased.	Quantity.		Cost.	
	Liters.	*Kilos.*	*Marks.*	*Pfennigs.*
Bread	....	7	1	35
Flour	....	½	....	18
Rye flour	....	1	....	28
Potatoes	8	....	....	35
Vegetables (peas, beans, rice, etc.)	....	¼	....	13
Chicory	....	⅛	....	5
Sugar	....	....	....	11
Rolls (bread)	....	....	....	12
Milk	½	....	....	6
Butter	....	¼	1	10
Sour-milk cheese	....	½	....	6
Meat	....	¼	....	30
Fish (herring)	....	....	....	8
Salt	....	¼	....	5
Pepper and other spice	....	....	....	2
Soap	....	....	....	10
Starch	....	....	....	5
Soda (for washing)	....	....	....	3
Petroleum	½	....	....	12
Candles	....	....	....	3
Total for 1 week	....	....	4	59
Total for 52 weeks	....	....	238	68

	Marks.[1]
6. House rent per year? *House is owned.*	
7. Clothing per year (shoemaker, tailor, linen, etc.)	10.00
8. Coal and wood per year	25.60
9. Dishes and cooking utensils per year	.50
10. School tax per year	.....
11. State tax per year	5.00

[1] The mark equals 24 cents, nearly.

	Marks.
12. Village tax per year	5.00
13. Insurance, life, accident, etc	.....
14. Expenditure for pleasure away from home, beer, tobacco, etc	.....
Total	46.10
Food, etc	238.68
Grand total	284.78

The topics of the remaining questions were the income derived from hand weaving and from other sources; whether any land was owned, if so, how it was worked; for whom weaving was done; what sort of cloth was made; the hours of work, and the expenditures necessary to carry on the weaving (weaver's paste, repairs of loom, etc.).

Among the further details reported are the following: The man and wife are spare and the man is often ill. The wife is healthy and in spite of her age can still weave. The man weighs 52 kilos (111.4 pounds), the woman 41 kilos (90 pounds), and neither is strong enough to work in the field. The wheat flour mentioned in the table was used to make weaver's paste. Vegetables mentioned are rice, millet, and grits. The milk used is skimmed milk or buttermilk.

The daily meals in this family are: Breakfast, flour soup (porridge), coffee, bread, butter; dinner, potatoes, salt, coffee, bread, butter; lunch (afternoon), bread, butter; supper, skimmed milk or buttermilk, bread, butter. This is the diet winter and summer, except that during the winter porridge is eaten for supper instead of milk.

In the opinion of the neighborhood this family lived very sparingly, yet it is the regular weaver's diet common to this region.

It will be seen that the greater part of this diet is bread, potatoes, flour, and coffee. The other food amounts to very little, in few cases exceeding 10 per cent of the total.

Green vegetables are too expensive to form any considerable factor in the diet except with families who own gardens.

The diet is such that it is impossible to improve it without material increase of cost. It is evidently shaped by long experience in the conditions in which the people are placed.

In these families no guests are entertained nor are any meals taken away from home except on an occasional feast day.

There is practically no food wasted. As one woman said in reply to a question concerning waste: "Everything is eaten up at our house. We are too poor to throw anything away."

These people say emphatically that they have enough to eat. They are completely satisfied after dinner and go to bed without hunger. Dr. von Rechenberg thinks this satiety is doubtless due in a great measure to the badly ventilated rooms in which they work and live, and to the monotony of their food.

The data collated in these dietary studies are concisely summarized in the following tables:

TABLE 27.—*Statistics of 28 families of hand weavers, numbers of persons, and annual income and expenditure of each family.*

Reference No. of family.	Family consists of[1]—						Total annual income.	Annual expenditures.		Ratio of expenditure for food to total expenditure.
	Father.		Mother.		Children.					
	Age.	Weight.	Age.	Weight.	Number.	Ages.		Total.	For food.	
	Years.	*Kilos.*	*Years.*	*Kilos.*		*Years.*	*Marks.*	*Marks.*	*Marks.*	*Per cent.*
2	56	52	71	41	0	....	286	286.28	212.16	74.1
3	55	72	53	38	0	....	445	473.88	366.60	77.4
9	34	59	32	54	0	....	478.64	457.34	299	65.4
24	53	....	51	....	0	....	321	308.66	231.92	75.1
37	57	55	56	50	0	....	344.1	386.1	234	63.6
38	57	75	56	67	0	....	319	319.46	219.96	68.9
45	59	60	53	53	0	....	405.6	385.80	314.08	81.5
1	20	50	24	55	4	1, 3, 9, 10	546	590.19	433.68	73.5
[2]4	43	61	40	57	2	12, 18	1307	1141.8	752.44	65.9
8	53	47	53	44	1	12	477.8	465.82	284.44	61.1
11	41	62	42	....	2	5, 6	435	439.36	317.72	72.3
12	44	56	40	52	4	3, 8, 11, 14	633	632.17	449.28	71.1
16	58	65	54	58	2	25, 3	469	484.53	366.08	75.5
17	58	61	47	....	1	6	465	468.17	340.08	72.6
21	57	62	38	50	4	$\frac{3}{4}$, 11, 13, 14	648	671	520	77.5
31	41	70	35	70	8	1, 3, 5, 7, 9, 11, 13, 15	676	707.6	580.16	83.3
32	48	50	48	53	3	10, 12, 14	603	618.	342.16	58.6
[2]33	50	44	49	59	2	8, 11	1115	1088.32	569.92	52.4
35	47	59	43	55	3	2, 4, 9	465	470	326.08	69.4
36	37	65	32	57	4	$\frac{1}{3}$, 3, 5, 6	465	464.14	342.16	73.7
40	31	57	28	49	2	3, 8	542	551	303.48	66
41	31	55	31	50	2	4, 8	298	297	195	65.7
43	36	50	34	41	2	3, 8	435	427.26	267.28	62.5
44	47	56	38	50	4	1, 5, 10, 13	470	462.42	268.84	58.1
46	28	52	23	....	2	2, 5	482	479.82	383.76	80
47	35	60	40	50	3	9, 10, 12	413.5	436.40	264.16	60.5
49	36	75	35	62	3	4, 7, 14	533.6	531.99	334.88	62.9
52	32	56	28	61	2	4, 6	320	338.32	248.04	73.3
Average	......	59		54	......	....	524	523	351	67
Average of childless families							395	397	268	68
Average of families with children							568	566	379	67

[1] Average weight of men and women, 57 kilos.
[2] Husband (in No. 4, the father and 18-year-old son) wove especially wide cloth.

TABLE 28.—*Weekly food consumption of families of weavers.*

	Reference No. of family.	Rye bread.	Rye flour.	Potatoes.	Butter.	Lard, etc.	Bacon, smoked.	Vegetables (dry).[1]	Rice.	Milk.	Skimmed milk.	Buttermilk.
		Kilos.	*Kilos.*	*Kilos.*	*Kilos.*	*Kilos.*	*Kilos.*	*Kilos.*	*Kilos.*	*Liters.*	*Liters.*	*Liters.*
Without children.	2	7	1	6	0.5				[2] 0.5		0.5	
	3	9	1	7.5	.75	0.25	0.25	1.5				
	9	9	.5	7.5	.75	.15		.25		1		
	24	7	1	5.25	.5				.25	.5	1.5	
	37	6.0	1	7.5	.5	.25		.25		1.5		
	38	7.5	.75	7.5	.5	.20		.25		1		1
	45	8.5	1.5	6	.5		.33	5			4	
With children.	1	15	3	12	1					.3		
	4	18	.75	20	1.5	.5			.5	.5		
	8	12	1.5	7.5	.5					.1		
	11	12	1	2.4	.75					.8		
	12	18	2	22.5	1	.25		1		1.5		
	16	12	1	9	.75	.25		.375			12.5	
	17	9		9	.75		.125				8	
	21	18	2	37.5	1.5					1		
	31	28	3	22.5	1	.5				.5		
	32	15	1.5	14.4	1	.15			.25	2		
	33	9	.5	22.5	1	.065	.375	.25		3.5		
	35	12	2	15	.75	.15		.25			1	
	36	10.5	1.25	18.75	.5	.375	1			2.5		
	40	13.25	1	25	.75					2		
	41	6	1.5	12	.35	.075		.25		.5		.7
	43	7	1	12	.5			.45			7	
	44	9	1.5	10	.5			.58			7	
	46	12	2	26.25	1					3		
	47	12.5	1	15	.5		.125		.25		1.5	
	49	12.5	1	10.6	1	.25			[3] .5	2		
	52	12	2	11.25	.5					.5		

	Reference No. of family.	Goat's milk.	Curd.	Eggs.	Herring.	Beef.	Pork.	Sugar.	Coffee.	Coffee substitutes, chicory, etc.	Rolls.	Sirup.	Fruit.
		Liters.	*Kilos.*	*No.*	*No.*	*Kilos.*	*Kilos.*	*Kilos.*	*Kilos.*	*Kilos.*	*Kilos.*	*Kilos.*	*Kilos.*
Without children.	2		0.5		1		0.25	0.15		0.125	0.3		
	3	1		1	2		.5	.125		.25	.9		
	9			1.5		0.55		.06	0.075	.15	.4		
	24		.25		1	.5		.125		.2			
	37					.25		.15	.10	.125			
	38		.375		1	.125		.03	.04	.075	.33		
	45					.5		.10	.10	.25	.66	0.5	
With children.	1		.3		2		.25	.07		.375	.9		
	4			6	6	1		.25	.5	.25	1.5		
	8		.5		1	.25		.125	.125	.25			
	11		.75		2					.25			
	12				2								
	16		.5		1	.125	.125	.125	.125	.125	.3		1
	17				.5	.5		.125	.125	.125	.5		.5
	21		.25						.04	.25			
	31		.47		1		.175	.25		.125	.83		
	32				.5	.25		.125		.375	.3		
	33		.235	2	2	1		.25	.25	.125	1.33		
	35		.55		2		.5	.05		.2	.67		
	36		.47		6	.325		.08	.06	.375	.9		
	40							.125	.125	.25			
	41		.47		2	.125	.125			.2			
	43		.25		1	.5		.15	.15	.15	1.4		
	44		1		1	.125		.05	.1	.125	.4		
	46		.47						.125	.25			
	47				.5	.09			.04	.125	.06		
	49				1	.5				.125	.83		
	52							.05		.075			

[1] I. e., as distinct from green vegetables. [2] Rice, barley grits, and millet. [3] Millet.

TABLE 29.—*Rates of potential energy of each food material to that of total food used by weavers.*

Kind of food.	Energy ratio to total energy.	Kind of food.	Energy ratio to total energy.
	Per cent.		*Per cent.*
Rye bread	54.8	Herring	.4
Rye flour	6.9	Beef and pork	.7
Potatoes	18.4	Sugar	.6
Butter	9.1	Coffee and coffee substitutes	.8
Fat and bacon	2.1	Rolls	1.6
Vegetables, rice, barley, grits, and millet	1.6	Sirup	.1
Milk, skimmed milk, buttermilk, and goat's milk	2.3	Fruit	.04
Curd	.6	Total	100
Eggs	.04		

TABLE 30.—*Estimated composition, digestibility, and potential energy of the food materials of weavers.*

[The figures are for estimated weights of nutrients in 100 grams of each kind of solid material, and for 100 cubic centimeters each of milk and beer. For eggs and herrings, however, the weights are for the number stated in each case.]

Food materials.	Nutrients.						Potential energy in nutrients.	
	Protein.		Fats.		[1]Carbohydrates.			
	Total.	Digestible.	Total.	Digestible.	Total.	Digestible.	Total.	Digestible.
Beef:	*Grams.*	*Grams.*	*Grams.*	*Grams.*	*Grams.*	*Grams.*	*Calories.*	*Calories.*
Medium fat, without bones	20.91	20.39	5.19	4.31	....	....	132	122
Medium fat, with 15 per cent bones	17.78	17.33	4.41	3.66	....	....	112	104
Pork:								
Fat, without bones	14.54	14.18	37.34	30.99	....	....	409	348
Fat, with 10 per cent bones	13.09	12.76	33.61	27.89	....	....	368	313
One herring, weight 135 grams, with 37 per cent refuse	16.07	15.67	14.36	11.92	....	....	199	174
Bacon, smoked	2.60	2.54	77.80	64.57	....	....	742	617
Lard	.26	.25	99.04	94.09	....	....	934	887
Cow's milk	3.51	3.35	3.77	3.66	4.96	4.96	69	67
Cow's milk, skimmed	3.22	3.04	.77	.74	4.91	4.91	40	39
Buttermilk	4.20	4.20	.96	.96	3.86	3.86	44	44
Curd	18.61	17.77	4.23	4.10	3.94	3.94	136	131
Butter	.71	.71	83.27	81.60	.58	.58	771	771
Goat's milk	3.62	3.46	4.05	3.93	4.52	4.52	70	68
One hen's egg, contents 47 grams	5.90	5.73	5.69	5.41	....	....	77	73
Rye bread	6.11	4.28	.43	.22	49.25	49.79	231	203
Rye flour	11.52	8.64	2.08	1.14	69.66	68.35	351	326
Wheat flour	11.82	9.46	1.36	.75	72.23	71.17	357	337
Wheat rolls	6.15	4.92	.44	.24	51.12	50.38	239	229
Potatoes, with 9 per cent refuse (skins, etc.)	1.80	[2].70	.14	.10	19.51	17.92	86	80
Rice	7.85	6.25	.88	.48	76.75	76.06	356	342
Millet	10.82	8.61	5.46	3	67.75	67.15	372	338
Barley grits	7.25	5.77	1.15	.63	76.19	75.50	353	340
Peas	22.85	18.85	1.79	.65	54.90	52.90	329	295
Beans	23.21	19.15	2.14	.77	56.20	54.20	340	303
Vegetables, dry, average	12.20	9.90	2.40	1.20	68.40	67.50	354	330
Fruit, fresh, average	.54	.54	....	....	[3]17	14.71	68	58
Sugar	....	....	....	....	99	99	[4]383	[4]383
Beer:								
Ordinary	....	....	[5].80	.80	1.63	1.63	[4]12	[4]12
Lager, light	....	....	[5]3.46	3.46	5.49	5.49	[4]45	[4]45
Bavarian export beer	....	....	[5]4.31	4.31	6.48	6.48	55	55
Coffee, infusion from 100 grams roasted coffee	....	....	5	5	13	13	95	95
Chicory and other coffee substitutes, infusion from 100 grams	....	....	1	1	55	55	221	221

[1] Omitting crude fiber except in potatoes and fruits. [2] The nonalbuminoid nitrogen is deducted. [3] Freeacid, 0.71; sugar, 7.97; cellulose, 4.32; nitrogen-free extract, 4.01; total, 17 per cent. [4] The energy of sugar is taken at 3,866 calories per gram. [5] Alcohol (instead of fats) in 100 cubic centimeters.

TABLE 31.—*Estimated nutrients and energy in daily food of each family of weavers.*

Reference No. of family.		Nutrients.						Potential energy in nutrients.	
		Protein.		Fats.		Carbohydrates			
		Total.	Digestible.	Total.	Digestible.	Total.	Digestible.	Total.	Digestible.
		Grams.	*Grams.*	*Grams.*	*Grams.*	*Grams.*	*Grams.*	*Calories.*	*Calories.*
Without children.	2	125	91	88	80	873	810	4,868	4,444
	3	170	124	203	183	1,200	1,119	7,462	6,810
	9	135	97	131	119	975	907	5,725	5,210
	24	127	91	78	71	813	756	4,534	4,129
	37	107	77	114	104	799	743	4,737	4,311
	38	117	84	109	99	1,029	957	5,158	4,694
	45	160	115	118	107	933	868	6,115	5,565
With children.	•1	252	179	175	159	1,817	1,688	10,036	9,155
	4	311	223	322	303	2,164	2,000	13,032	11,966
	8	170	126	86	78	1,253	1,155	6,613	5,991
	11	212	153	118	107	1,643	1,528	8,597	7,823
	12	261	188	188	171	2,105	1,958	11,347	10,320
	16	229	165	164	149	1,372	1,276	8,028	7,305
	17	168	121	126	115	1,056	982	6,120	5,569
	21	300	216	210	191	2,541	2,363	13,345	12,144
	31	381	274	237	216	2,980	2,771	15,998	14,558
	32	217	156	171	156	1,717	1,597	9,441	8,591
	33[1]	219	158	217	197	1,508	1,402	8,991	8,182
	35	192	138	161	147	1,568	1,458	8,733	7,947
	36	216	156	170	155	1,516	1,410	8,590	7,817
	40	207	149	117	106	1,783	1,658	9,133	8,311
	41	142	102	81	74	956	889	5,217	4,747
	43	188	135	88	80	1,151	1,070	6,269	5,705
	44	212	153	92	84	1,216	1,131	6,675	6,074
	46	219	158	158	144	1,822	1,694	9,772	8,893
	47	179	129	91	83	1,450	1,349	7,400	6,789
	49	194	140	188	171	1,408	1,309	8,258	7,515
	52	170	122	78	71	1,374	1,278	6,996	6,366

[1] Husband (in No. 4, father and 18-year-old son) wove extra wide cloth.

TABLE 32.—*Estimated amounts of protein fats and carbohydrates and potential energy of the digestible food of the men and the women of the weavers' families, per day.*

Reference No. of family.		Men.				Women.				Potential energy for adults.
		Protein.	Fats.	Carbohydrates.	Potential energy.	Protein.	Fats.	Carbohydrates.	Potential energy.	
		Grams.	*Grams.*	*Grams.*	*Cals.*	*Grams.*	*Grams.*	*Grams.*	*Calories.*	*Calories.*
Without children.	2	49	43	437	3,297	42	37	373	2,047	2,672
	3	81	114	698	4,250	43	69	421	2,560	3,355
	9	50	61	467	2,682	47	58	440	2,528	2,620
	24	47	37	419	2,125	44	34	394	2,004	....
	37	37	51	361	2,093	40	53	382	2,218	2,142
	38	44	51	496	2,435	40	48	401	2,259	2,029
	45	63	59	477	3,057	52	48	391	2,508	2,634
With children.	1	43	43	461	2,503	52	40	492	2,668	2,733
	4	69	95	622	3,719	60	82	540	3,240	3,240
	8	49	30	455	2,359	47	29	435	2,257	2,628
	11	56	39	558	2,862	51	36	509	2,600	....
	12	47	43	496	2,611	45	41	470	2,475	2,631..
	16	53	47	404	2,317	49	44	375	2,149	....
	17	52	49	419	2,378	48	45	386	2,192	....
	21	57	50	616	3,162	49	43	533	2,738	2,991
	31	47	37	474	2,489	47	37	474	2,480	2,170
	32	40	40	413	2,267	42	42	429	2,308	2,476
	33	51	66	466	2,686	55	69	487	2,866	2,800
	35	45	47	471	2,508	43	45	449	2,450	2,509
	36	47	47	434	2,369	58	53	470	2,692	2,171
	40	55	40	618	3,098	50	36	559	2,801	3,098
	41	37	27	322	1,717	35	25	302	1,611	1,759
	43	51	30	405	2,160	45	26	354	1,891	2,359
	44	42	23	311	1,666	39	21	288	1,544	1,686
	46	57	53	615	3,300	59	54	631	3,385	....
	47	38	25	401	2,019	34	22	355	1,787	1,950
	49	42	51	393	2,256	37	45	346	1,986	1,879
	52	42	25	439	2,180	44	20	405	2,317	2,204

Estimated average daily metabolism of material and energy per average person.

	Protein.	Fats.	Carbo-hydrates.	Potential energy.
	Grams.	*Grams.*	*Grams.*	*Calories.*
Total food consumed	65	49	485	2,703
Digestible portion of food consumed	47	45	451	2,461

TABLE 33.—*Estimated potential energy per day in the food (digestible portion) of the children of weavers' families.*

[Ages of children.]

Reference No. of family.	Months.		Years.							
	Four.	Nine.	One.	Two.	Three.	Four.	Five.	Six.	Seven.	Eight.
	Cals.	*Cals.*	*Cals.*	*Cals.*	*Cals.*	*Cals*	*Cals.*	*Cals.*	*Cals.*	*Cals.*
1			564		915					
4										
8										
11							1,157	1,196		1,176
12					896					
16					790					
17								999		
21		643								
31			543		881		1,053		1,120	
32										
33										
35				755		973				1,212
36	521				804		961	992		
40					1,044					1,369
41						646				772
43					717					940
44			350				679			
46				993			1,414			
47										
49						823			954	
52						873		983	983	
Average	521	643	486	874	864	829	1,053	1,043	1,037	1,094

Reference No. of family.	Years.								
	Nine.	Ten.	Eleven.	Twelve.	Thirteen.	Fourteen.	Fifteen.	Eighteen.	Twenty-five.
	Cals.	*Cals.*	*Cals.*	*Cals.*	*Cals.*	*Cals.*	*Cals.*	*Cals.*	*Cals.*
1	1,236	1,272							
4				1,975				3,035	
8				1,376					
11									
12			1,378			1,789			
16									2,048
17									
21			1,605		1,918	2,083			
31	1,190		1,354	1,364	1,618		1.819		
32		1,126				1,615			
33			1,420						
35	1,199								
36									
40									
41									
43									
44		790			1,043				
46									
47	910	936		1,134					
49									
52						1,497			
Average	1,134	1,031	1,439	1,462	1,526	1,746	1,819	3,035	2,048

TABLE 34.—*Estimated money and fuel values of food materials in diet of weavers.*

Food materials.	Market price.	Fuel value of digestible portion.a	Calories (net values) obtained for one mark.	1,000 calories (net values) cost.
		Calories.	*Calories.*	*Marks.*
Potatoes, 9 per cent refuse	0.056 mark per kilo	800	14,288	0.070
Rye flour	.2467 mark per kilo	3,260	13,213	.076
Rye bread	.1867 mark per kilo	2,030	10,873	.092
Wheat flour	.3256 mark per kilo	3,370	10,346	.097
Vegetables, "dry" b	.346 mark per kilo	3,300	9,538	.105
Fruit, fresh	.08 mark per kilo	580	7,250	.138
Skimmed and buttermilk	.06 mark per liter	420	7,014	.143
Roll	.333 mark per kilo	2,290	6,877	.145
Lard, rendered	1.35 mark per kilo	8,870	6,473	.152
Curd	.213 mark per kilo	1,310	6,157	.163
Cow's milk	.123 mark per liter	670	5,447	.184
Bacon, smoked	1.49 mark per kilo	6,170	4,140	.241
Sugar	.95 mark per kilo	3,830	4,043	.248
Butter	2.06 mark per kilo	7,560	3,667	.272
Pork, with 10 per cent bones	1.146 mark per kilo	3,480	2,729	.301
Coffee, artificial	.80 mark per kilo	2,210	2,563	.362
Herring, weight 135 grams, 50 per cent refuse	.08 mark per piece	174	2,175	.460
Eggs, contents 47 grams	.05 mark per piece	73	1,440	.685
Beef, with 15 per cent bones	1.08 mark per kilo	1,220	963	.885
Coffee, genuine, roasted	2 marks per kilo	950	475	2.105

a In one kilo, liter, or piece. b Rice, millet, grits, barley, legumes.

The value of this study of Von Rechenberg's, so far as the accuracy of its results is concerned, is to be judged from two standpoints, namely, the reliability of the statistical information, including especially that regarding food consumption, and the accuracy of the values used for the calculation of composition, digestibility, and potential energy of the food materials.

Concerning the accuracy of the statistical information, it is to be noted that the data were not collected by Von Rechenberg in person, but were furnished by the people whom he was studying.

Von Rechenberg's work, like nearly all thus far done, falls below the ideal in still another respect. No determinations were made of either the composition or the digestibility or the fuel value of the food. All the figures for composition are taken from the compilations of König[1] and other standard authorities. These values are doubtless as close to the facts as are possible without actual analyses.

These estimates of quantities of food, and of nutrients, digestibility and potential energy of the food consumed by the people in Zittau, which are carried out in great detail remind one of the method of estimating distances, which consists in measuring feet by the eye, inches by a foot rule, and fractions of an inch by a micrometer screw, and taking the sum of the figures thus obtained for the actual measure. Nevertheless such computations have their value and are referred to here, not merely as an illustration of the ways in which an investigator who had devoted a great deal of attention to the study of the fuel value of food materials seeks to apply his results, but also as an indication of some of the ways in which accuracy must be sought in future research. When it shall become possible to make experiments in which the income and

[1] Chemie der Nahrungs- und Genussmittel, second edition.

outgo shall be accurately weighed, measured, and analyzed, such refinements of computation as those employed by Von Rechenberg may actually come into play.

DIETARIES OF PEOPLE OF THE POORER CLASSES IN NAPLES.

A careful and somewhat extended study of the dietaries of people of the poorer classes in Naples has been recently made by Manfredi.[1] An excellent feature of this investigation is found in the determination of not only the composition of the food consumed, but also the proportions actually digested. In the following account, which is greatly condensed from the original, a number of evident errors in the latter are corrected.

Manfredi is a native Neapolitan and could easily reach the people he desired to study, namely, the small wage earners, the beggars "lazzaroni," and in general those who live from hand to mouth.

Little of the food consumed by this class of people is cooked at home. Small shops—bettole—abound, in which the food is purchased ready cooked. This does not mean that the best food of the kind is sold for a low price as in the German "people's kitchen" (*Volksküche*). The food is really of the poorest quality and the price is relatively high, a meal costing usually from 5 to 30 centesimi (1 to 6 cents). The food of these people contains a larger variety of articles than is usual with the poor classes in Europe. The variety consists in the number of kinds of macaroni and other pastes, and the green vegetables which grow so abundantly in the neighborhood of Naples. Macaroni and bread are, perhaps, the two most important foods.

Macaroni and other forms of flour paste are to the Neapolitan what polenta is to the people of northern Italy and rice is to some of the people of Asia. Sometimes the dough is pressed in flat cakes baked with fat or cheese and tomatoes or with small fresh fish; this cake is called "pizza." Indeed, numerous dishes of this sort are in common use. The bread is almost always white—black bread is rarely seen—and often as much as 1 to 1½ kilos per day is eaten by one person. In times of distress almost the only food is bread. Beef is too expensive to form any considerable part of the diet of the poor. When meat is eaten it is usually mutton in summer and pork in winter. Much oftener, however, the heart, lungs, liver, brains, and the parts which we ordinarily count as refuse are eaten, and dishes in great variety are made from them. It is curious that while fish abound in the waters near Naples, very little is eaten fresh. Sometimes small fishes are seasoned in various ways and fried in fat. On the other hand large quantities of dried and salted fish are eaten, chiefly cod and hake, (gadus morrhua and gadus melucius).[2] These are either fried or made into a kind of soup with tomatoes or cooked with oil and vinegar. Milk and eggs are too expensive to be eaten much, and cheese is used

[1] Arch. Hyg., 17, 1893, p. 552.

[2] *Gadus melucius* (?)—*G. Merluccius*, L., *Merluccius vulgaris*, Günther.

more as a condiment than as an article of food. Rice is too dear to be much employed, but potatoes are abundant especially in summer. Dry beans are used in large and peas in small quantities nearly always for soups. The garden vegetables which are so abundant are cabbage, cauliflower, green salad, tomatoes, carrots, onions, etc. In summer enormous quantities of fruit are eaten, so much that it sometimes seems as if the beggars must live on fruit alone, which of course is not the case. The drink which is almost always used at meal times is a light wine, often so much diluted that it could hardly be called wine.

The author's description of the people and the methods of observation is in substance as follows: Dietaries of 8 individuals were studied. The studies extended from January to July, thus making it possible to observe the effect of different seasons of the year upon the food consumption, a by no means unimportant factor in Naples. The individuals were from the classes of the very poor that make up a large part of the population of Naples. They included (1) workmen of the lower class who prefer light and intermittent labor, such as cobblers, carpenters, masons, and (2) people who are without regular employment and much of the time idle, the wandering creatures who pass the day on the street and live on what chance may bring, such as messengers, peddlers from house to house, and beggars, lazzaroni.

During the investigation each individual was very carefully watched. Every precaution was taken to keep them in the laboratory day and night or to have them under surveillance if they went outside, and to make sure they associated only with persons connected with the laboratory and did not come in contact with ordinary companions. In so far as possible each person was provided with occupation of the kind and amount to which he was accustomed. For instance, the mason mended the tile roof of the laboratory; the carpenter worked with a carpenter employed in the Hygienic Institute at that time; old shoes were given to the cobbler to mend, others were employed in various ways in cleaning the laboratory, and even the beggars "who were accustomed to do nothing were provided with their usual employment, which was not a difficult matter." The peddlers, who were accustomed to considerable exercise, found this in climbing the many stairs and walking in the garden of the institute; they were sometimes allowed to walk in the city, being accompanied by a member of the laboratory staff or by Manfredi himself.

In the supply of food the usage of families of the poorer classes in Naples was followed, in that two meals were eaten per day, a hearty one at noon and a light, sometimes a very light, one at night. In some cases, however, the order was reversed, the heartier meal being eaten at night. It should be said that such regularity in eating is by no means universal and rigidly adhered to among the poor people. Very many, particularly those leading a vagrant life, are accustomed to long fasts, followed by eating to excess. Others, particularly women, have no definite meal times, but nibble at all hours of the day. The latter custom, however, was not permitted in these investigations, both because it is the exception rather than the rule, and because it would have made the number of analyses very great, and thus have increased the chances of error. When the individual could be trusted his satiety was taken as the measure of the amount of food he required. Sometimes the amount was determined from the size of the portions of cooked food furnished and from information obtained at the cookshop where they were accustomed to obtain their food. According to the habit of the individual the midday meal was either prepared at home or purchased ready cooked from a cookshop. In every case one and one-half or two portions were provided, one for the subject and the remainder for analysis. The period of observation for each person was from 3 to 7 days. Each day was reckoned from 8 o'clock in the morning to 8 the next morning. The food between these two hours was regarded as the food for the day.

Specimens of the food of each meal were taken for analysis. The corresponding urine and feces of 24 hours were likewise taken for analysis. It was considered that the portion of the former which belonged to the day of the experiment would be voided during the 24 hours beginning 12 hours after the beginning of the day, i. e., as the experimental day began at 8 a. m. the urine for that day would include what was collected during the 24 hours beginning with 8 p. m. The separation of feces was made with the aid of seed of grapes taken with the last meal before and the first meal after the experimental period. A fasting period of 16 hours was allowed to intervene between the first meal of the experiment and the last preceding, and the same period of fast intervened between the last meal of the experiment and the next succeeding one. With the simple diet of vegetable food to which the people were accustomed the separation was very easy. The analysis of food and excreta were made by the usual methods.

The following table, which is translated literally from the original, shows the statistics gathered in one of the eight cases. Like statistics are given for each of the seven other cases.

TABLE 35.—*Dietary statistics of a cobbler.*

Felix Candelmo, from Cimitile (Nola), lived in Naples since childhood, 34 years old. Works when he has anything to do, at other times is idle. Constitution weak, poorly nourished, muscular system little developed. Weight on the 5 days of the investigation, 55 to 55.5 kilos (121 to 122.1 pounds); height, 1.66 meters (5 feet 4.5 inches); breast measure, 78 centimeters (30.7 inches); dynamometer right hand, 95. Worked in laboratory daily on an average 6 to 7 hours. Midday meal cooked at home.

Date.	Food consumed.	Liquid.	Water-free substance.	Nitrogen.	Protein.	Ether extract.	Carbohydrates and ash.	Feces.						Urine.				Total excreted nitrogen.
								Liquid.	Water-free substance.	Nitrogen.	Protein.	Ether extract.	Carbohydrates and ash.	Quantity.	Specific gravity.	Nitrogen.	Protein.	
		Grams	*Grams*	*Grams*	*Grams*	*Grams*	*Grams.*	*Grams*	*Grams*	*Grams*	*Grams*	*Grams*	*Grams*	*Grams*	*Grams*	*Grams*	*Grams*	*Grams*
1892. Feb. 24.	Dinner: Bean soup, 482; bread, 200	682	403	7	43.75	17.9	341.4											
	Supper: Fried hake, 45; bread, 152	197	42.7	1.34	8.37	3.8	30.5											
	(Price, 40 centesimi.) Total	879	445.7	8.34	52.12	21.7	371.9											
Feb. 25.	Dinner: Cow's foot soup with potatoes, 1,245; bread, 275	1,520	386.6	9.28	58	21.8	306.8											
	Supper: Dry figs, 150; bread, 190	340	90.3	1.20	7.50	.6	82.2											
	(Price, 40 centesimi.) Total	1,860	476.9	10.48	65.50	22.4	389	245	25.8	1.30	8.12	5.9	14.8	658	1,022	7.80	48.75	9.10
Feb. 26.	Dinner: Soup with hake, 445; bread, 197	642	234.1	11.06	69.12	23.6	141.4											
	Supper: Baked pizza with anchovies, 97	97	58.9	.94	5.87	5.3	47.7											
	(Price, 55 centesimi.) Total	739	293	12	74.99	28.9	189.1	258	26.7	1.09	6.81	5.0	14.9	970	1,026	10.08	63	11.17
Feb. 27.	Dinner: Cauliflower soup with beef liver, etc., 745; bread, 200	945	310	8.10	50.62	18.9	240.5											
	Supper: Cheese, 54; bread, 268	322	207.9	5.75	35.93	6.9	165.1											
	(Price, 55 centesimi.) Total	1,267	517.9	13.85	86.55	25.8	405.6	190	23	.87	5.43	3.1	14.5	940	1,023	8.65	54.06	9.52
Feb. 28.	Dinner: Macaroni with tomato, 558; meat croquettes, 115; bread, 210	883	415.4	11.16	69.75	40.6	305.1											
	Supper: Olives, 52; bread, 163	215	99.9	1.49	9.31	7.5	83.1											
	(Price, 65 centesimi.) Total	1,098	515.3	12.65	79.06	48.1	388.2	218	21.2	1.31	8.19	2.7	13.3	695	1,025	11.90	74.37	13.21
Feb. 29.								286	34.4	2	12.50	4.9	17	1,086	1,020	8.82	55.12	10.82
Mar. 1.								192	31.4	1.99	12.43	2.7	16.3					1.99
	Total	5,837	2,248.8	57.32	358.18	146.9	1,743.8	1,389	165.5	8.56	53.48	24.3	90.8			47.25	295.30	55.81
	Daily average	1,167.4	449.7	11.46	71.62	29.4	348.8	278	33.1	1.71	10.68	4.8	18.16			9.45	59.06	11.16

The results of the 8 investigations are summed up in the following table:

TABLE 36.—*Summary of statistics of eight dietaries of poor people in Naples.*

Subject.	Food consumed.								
	Weight.	Number of days.	Liquid.	Water-free substance.	Nitrogen.	Protein.	Ether extract.	Carbohydrates and ash.	Sodium chlorid.
	Kilos.		*Grams.*	*Grams.*	*Grams.*	*Grams.*	*Grams.*	*Grams.*	*Grams.*
Cobbler	55	5	1,167.4	449.7	11.46	71.62	29.4	348.8	....
Cobbler	47	5	1,311.6	532.9	12.65	79.06	45.6	408.6	10.8
Old woman, servant	38	3	976.6	412.4	9.31	58.18	19.8	334.4	6.3
Carpenter	62	7	1,377.3	624.9	15	93.75	56.2	475.1	9.4
Young woman, servant	48	4	1,179.2	425.4	10.00	62.87	19.9	342.8	11.4
Mason	55	5	1,148.8	489.6	11.28	70.49	28.5	390.6	....
Beggar	50	4	1,271	447.8	10.45	65.28	28.2	354.2	10.8
Peddler (woman)	52	5	954	385.8	9.72	60.75	27.9	297.2	....
Maximum	....	......	1,377.3	624.9	15	93.75	56.2	475.1	11.4
Minimum	....	......	954	385.8	9.31	58.18	19.9	207.2	6.3
Average	51	......	1,173.2	471.1	11.24	70.25	31.9	368.9	9.7

Subject.	Feces.							Urine.			Total excreted nitrogen.
	Liquid.	Water-free substance.	Nitrogen.	Protein.	Ether extract.	Carbohydrates and ash.	Sodium chlorid.	Nitrogen.	Protein.	Sodium chlorid.	
	Gms.	*Gms.*	*Gms.*	*Gms.*	*Gms.*	*Gms.*	*Gms.*	*Gms.*	*Gms.*	*Gms.*	*Gms.*
Cobbler	278	33.7	1.71	10.71	4.8	18.6	......	9.45	59.06	......	11.16
Cobbler	141.6	33.1	2.19	13.68	7.1	12.3	1.3	10.56	66	9.6	12.75
Old woman, servant	116	21.6	1.32	8.28	1.5	11.7	1.5	7.69	48.08	4.6	9.01
Carpenter	245.9	45.7	3.06	19.09	5.7	20.8	1.1	11.62	72.62	9	14.68
Young woman, servant	171.7	39.4	3.14	19.62	3.8	16	.4	6.84	42.75	9.9	9.98
Mason	177.4	38.6	2.97	18.56	2.7	18.4	......	8.04	50.25	......	11.01
Beggar	113	22	1.405	8.78	2.1	11.3	.6	9.245	57.78	10.8	10.65
Peddler (woman)	153	24.8	1.57	9.81	3.8	11.1	......	8.05	50.31	......	9.62
Maximum	278	45.7	3.14	19.62	7.1	20.8	1.5	11.62	72.62	10.8	14.68
Minimum	113	21.6	1.32	8.28	1.5	11.1	.4	6.84	42.75	4.6	9.01
Average	178.3	32	2.17	13.50	3.9	15	1	8.937	55.85	8.8	9.15

TABULAR STATEMENTS OF RESULTS OF DIETARY STUDIES.

In Table 37, which follows, the final results of the 138 selected studies of dietaries, mentioned on page 144 above, are recapitulated. It was thought desirable to classify these dietaries by countries rather than by classes of people and occupation.

TABLE 37.—*Foreign and American dietaries.*

[Quantities per man per day.]*

Reference No.	Dietaries.	Number of persons.	Nutrients. Protein.	Fats.	Carbohydrates.	Potential energy.	Nutritive ratio.
	FOREIGN. (EUROPEAN, JAPANESE, AND JAVANESE.)						
	ENGLAND.		*Grams.*	*Grams.*	*Grams.*	*Calories*	1:
1	Well-fed tailors (prisoners)	2	131	39	525	3,055	4.7
2	Hard-worked weavers (prisoners)	2	151	43	622	3,570	4.8
3	Royal engineers, at active work	495	144	83	631	3,950	5.7
	DENMARK.						
4	Physician, 37 years old, Copenhagen	1	135	140	250	2,880	4.2
5	Principal of school, 35 years old, wife of No. 4	*1	95	107	220	2,285	4.9
	SWEDEN.						
6	Medical students, 22 to 27 years of age, Stockholm	5	127	114	300	3,035	4.8
7	Mechanics, etc., at moderate work, southern Sweden, average of 6 individual dietaries	6	134	79	[1]523	3,435	5.2
8	Mechanics, etc., at severe work, southern Sweden, average of 5 individual dietaries	5	189	110	[2]714	4,725	5.1
	RUSSIA.						
9	Factory operatives, near Moscow, average of 9 families	(*)	106	49	488	2,890	5.7
10	Factory operatives, near Moscow, average of 50 boarding clubs composed of men	[3]2,670	132	80	584	3,680	5.8
11	Factory operatives, near Moscow, average of 10 boarding clubs composed of women and boys.	*[3]234	98	51	487	2,875	6.1
12	Peasants near Moscow, average of men's dietaries	......	129	33	589	3,250	5.1
13	Peasants near Moscow, average of women's dietaries	(*)	102	28	471	2,610	5.2
	GERMANY.						
	1. Scantily nourished.						
14	Laborer's family, father, mother, and child 6 years old	3	52	32	287	1,690	6.9
	WORKING PEOPLE IN LEIPSIC, SAXONY.						
15	Girl in factory, wages $1.20 (5 marks) per week, pale and sickly, food mostly bread and potatoes, half a pound of meat per week, cost of food 8.2 cents per day	*2	52	33	301	1,940	8.1
16	Girl in printing office, wages $1.20 per week, food similar to preceding, cost 10.3 cents per day	*2	65	39	303	1,870	6
17	Widow, straw plaiter, with 5 children, food mostly bread and potatoes, occasionally meat, cost of food 9.6 cents per day	*6	72	56	440	2,620	7.9
18	Two girls, one 24 years old, in paper factory, wages $2.16 (9 marks), the other 21 years, seamstress, wages $1.92 (8 marks), cost of food 8.6 cents per day	*3	56	51	229	1,645	6.2
19	Girl, 18 years old, in book-bindery, wages $2.16 per week, cost of food 8.6 cents per day	*1	61	41	347	2,055	7.2
20	Painter, wages $4.32 per week, unmarried, cost of food 11.5 cents per day	1	87	64	366	2,450	5.9
21	Cabinetmaker, with wife and 6 children, wages $1.92 per week, cost of food 11 cents per day	8	77	57	466	2,760	7.7
22	Druggist's clerk, with wife and 2 children, annual income $283 (1,180 marks), very little physical exercise, cost of food 11.3 cents per day	*4	71	69	351	2,370	7.1
23	Farm laborer near Leipsic, with wife and 4 children, food mainly vegetable, cost of food 7.2 cents per day	6	80	37	504	2,740	7.4

* Quantities per person per day.
[1] Including 22 grams alcohol. 1 gram alcohol = 1.71 grams carbohydrates.
[2] Including 24.2 grams alcohol.
[3] An approximate estimate.

TABLE 37.—*Foreign and American dietaries*—Continued.

Reference No.	Dietaries.	Number of persons.	Nutrients. Protien.	Fats.	Carbohydrates.	Potential energy.	Nutritive ratio.
	WORKING PEOPLE IN LEIPSIC, SAXONY—cont'd.						
			Grams.	*Grams.*	*Grams.*	*Calories*	*1:*
24	Farm laborer, Prussia, very poor, diet mostly vegetable	3	83	17	573	2,845	7.4
25	Laboring woman of poorer class, Munich, vigorous, but at rather hard work, food hardly adequate	*1	76	23	334	1,895	5.1
26	Hand weavers, Zittau, Saxony, average of adults in 28 families	1*56	65	49	485	2,705	9.2
	Average, Nos. 14 to 26	*96	69	45	381	2,275	7
	2. Prisoners.						
27	Inmates of Badstrasse prison in Munich, doing no work	(*)	87	22	305	1,810	4.1
28	Inmates of prison in Brandenburg, without work	(*)	109	34	574	3,115	6
29	Inmates of house of correction in Munich, at work	......	104	38	521	2,915	5.8
30	Inmates of house of correction in Brandenburg, at work	......	127	29	639	3,410	5.5
	3. People not having active exercise.						
31	Inmates of house for old women in Munich (Pfründneranstalt)	(*)	80	49	266	1,875	4.7
32	Inmates of house for old men and women in Munich (Pfründneranstalt)	*477	92	45	332	2,155	4.7
33	Man at rest, average of 3 tests in respiration apparatus	1	137	72	352	2,675	3.8
34	Physician, 31 years old (Professor Beneke), Oldenburg, diet not quite sufficient to maintain bodily condition	1	90	79	285	2,270	5.2
35	Same, diet estimated sufficient for normal maintenance	1	94	109	284	2,565	5.6
36	Professor Ranke, Munich, in respiration apparatus	1	100	100	240	2,325	4.7
37	Same, food slightly different from No. 36	1	126	85	213	2,180	3.2
38	Lawyer, Munich	3	80	125	222	2,400	6.3
39	Young physician, Munich	1	127	89	362	2,835	4.4
40	Young physician, Munich	1	134	102	292	2,695	3.9
	Average of Nos. 35 to 40, 6 dietaries of well-to-do professional men, food largely animal	8	110	102	269	2,505	4.5
41	Official in civil service, a vegetarian	1	74	58	490	2,850	8.4
	4. People with more or less active muscular work.						
42	Mechanic in comfortable circumstances at light work, 60 years old, Munich	1	117	68	345	2,525	4.3
43	Shoemaker, Rostock	1	108	77	378	2,710	5.1
44	Upholsterer, a strict vegetarian, 28 years old, Munich	1	54	22	573	2,775	11.5
45	Well-paid mechanics, Munich	3	151	54	479	3,085	4
46	Carpenters, coopers, locksmiths, Bavaria, average of 11 dietaries	......	122	34	570	3,150	5.3
47	Porter, Munich, 36 years old, unmarried	1	133	95	422	3,160	4.8
48	Cabinetmaker, 40 years old, Munich	1	131	68	494	3,195	5
	Bavarian mechanics, etc., at moderate work, average of Nos. 45 to 48	......	134	63	491	3,150	4.7
49	Farm laborers, Bavaria, average of 5 dietaries	......	137	55	542	3,295	4.9
50	Miners at severe work, Prussia	4	133	113	634	4,195	6.7
51	Brewery laborers, Bavaria, average of 5 dietaries	......	149	61	755	4,275	6
52	Brickmakers (contract laborers from Italy) near Munich; diet, maize meal and cheese	300	150	94	683	4,290	6
53	Machinists, etc., Krupp gun works, Essen	800	139	113	677	4,395	6.7

* Quantities per person per day.

[1] This average is reduced to terms corresponding to a person weighing 57 kilos (125 pounds), the average weight of all the men and women of the families.

TABLE 37.—*Foreign and American dietaries*—Continued.

Reference No.	Dietaries.	Number of persons.	Nutrients. Protein.	Fats.	Carbohydrates.	Potential energy.	Nutritive ratio.
	WORKING PEOPLE IN LEIPSIC, SAXONY—cont'd.						
	4. *People with more or less active muscular work*—Continued.						
			Grams.	*Grams.*	*Grams.*	*Calories*	1:
54	Farm laborers, Bavarian highlands, large, muscular men at severe work, average of 3 dietaries	3	137	202	546	4,680	7.3
55	Lumbermen, Bavarian highlands, large, muscular, vigorous, doing heavy work in rigorous climate, average of 3 dietaries	3	130	292	724	6,215	10.7
56	German army ration, peace footing		114	39	480	2,800	5
57	German army ordinary ration, war footing		134	58	489	3,095	4.6
58	German army extraordinary ration, in war		192	45	678	3,985	4.1
59	German army extraordinary ration, Franco-German war		157	285	331	4,650	6.2
	AUSTRIA.						
60	Farm laborers, Transylvania, at harvesting; diet, maize meal and beans	15	159	62	977	5,235	7
	ITALY.						
	Poor people in Naples, lower class.						
61	Mechanics, etc., average of 5 dietaries	5	76	38	396	2,290	6.3
62	Servants, etc. (women), average of 5 dietaries	*3	61	23	325	1,795	6.2
63	Peasant, farm laborer, near Ferrara, average of winter and summer dietaries	1	118	65	628	3,665	6.6
64	Wife of No. 63, average of winter and summer dietaries	*1	95	54	499	2,940	6.5
65	Son of Nos. 63 and 64, 14 years old, average of winter and summer dietaries	*1	67	39	332	2,000	6.3
66	Italian army ration, peace footing		114	14	592	3,025	5.5
	FRANCE.						
67	Physician, 48 years old		92	61	235	1,910	4.1
	CHILDREN (GERMANY, RUSSIA, AND SWITZERLAND).						
68	Under 2 years old, average of 3 individual dietaries	*3	27	20	115	770	5.9
69	From 2 to 6 years old, average of 14 individual dietaries	*14	53	42	155	1,245	4.7
70	From 6 to 10 years old, average of 8 individual dietaries	*8	65	45	217	1,575	4.9
71	From 10 to 14 years old, average of 13 individual dietaries	*13	72	50	249	1,780	5
72	Children, 6 to 15 years old, in orphan asylum, Munich	(*)	79	37	247	1,680	4.2
73	Boys, 8 to 15 years old, in children's home at Gehlsdorf, near Rostock	*38	87	50	508	2,905	7.1
74	Girls, 14 to 19 years old, Krupp industrial school, Essen	*71	101	75	415	2,815	5.8
	JAPAN.						
75	Prisoners without work, Tokyo	(*)	48	7	372	1,785	8.1
76	Prisoners at work, Tokyo, average of 2 dietaries	(*)	66	9	544	2,585	8.5
77	Employees in retail store, Tokyo	*72-85	55	6	394	1,895	7.4
78	Y. Mori, assistant in University of Tokyo; dietary of self (ordinary diet with meat)	1	123	21	416	2,405	3.8
79	Students, Kioto, average of 2 individual dietaries	2	98	16	438	2,345	4.8
80	Cadet school at Tokyo	(*)	83	14	631	3,060	8
81	Government school, Tokyo; pupils, 17 to 25 years old	*130	115	30	635	3,355	6.1
82	Private school in Tokyo; pupils, 11 to 21 years old	*21	79	13	470	2,370	6.3
	JAVA.						
83	Java village, World's Fair, Chicago	5	66	19	254	1,490	4.5

* Quantities per person per day.

TABLE 37.—*Foreign and American dietaries*—Continued.

Reference No.	Dietaries.	Number of persons.	Nutrients.			Potential energy.	Nutritive ratio.
			Protein.	Fats.	Carbohydrates.		
	AMERICAN (UNITED STATES AND CANADA).[1]						
	SERIES A.—MISCELLANEOUS, FACTORY OPERATIVES, MECHANICS, ETC., MASSACHUSETTS.						
	Boarders, operatives in cotton mills:		*Grams.*	*Grams.*	*Grams.*	*Calories*	1:
84	Boarding house, Lowell	77	132	227	545	4,890	8
85	Boarding house, Lowell	70	132	200	594	4,670	7.6
86	Boarding house, Lowell	150	105	136	477	3,650	7.5
87	Boarding house, Lynn, shoe factory operatives, dressmakers, clerks, etc	36	114	150	522	4,000	7.5
88	Boarding house, Lawrence, mill operatives	80	127	195	523	4,480	7.6
89	Family, East Cambridge, father glass blower	6	95	132	481	3,590	8.2
90	Family, Boston, husband, machinist	2	182	254	617	5,640	7.6
	Average	421	127	185	531	4,415	7.5
91	Boarding house, Boston, teamsters, marble workers, etc., at severe work	12	254	363	826	7,805	6.5
92	Boarding house, brickmakers, at severe work	237	180	365	1,150	8,850	11
	Average	670	147	225	632	5,285	7.8
93	French Canadian families, in North Cambridge, fathers, brickmakers, at severe work	4	132	236	750	5,810	9.7
94		5	100	173	545	4,250	9.4
95		5	95	145	514	3,845	8.9
	Average	14	109	185	603	4,635	9.3
	SERIES B.—FRENCH CANADIANS, FACTORY OPERATIVES, MECHANICS, ETC., MASSACHUSETTS.						
96	Family, Lawrence, mill operatives	6	114	186	522	4,340	8.3
97	Boarding house, Holyoke, operatives in paper mills	18	95	268	327	4,220	9.8
98	Boarding house, Holyoke, factory operatives	10	141	268	554	5,340	8.2
99	Family, Holyoke, mill operatives	6	82	132	495	3,595	9.7
100	Family, Worcester, mill operatives	10	114	127	504	3,715	6.9
101	Family, Worcester, father, printer	4	118	177	509	4,215	7.7
102	Family, Lowell, blacksmiths at hard work	3	200	304	795	6,905	7.4
	Average	57	123	209	529	4,620	8.2
	Average	71	119	202	552	4,625	8.5
	SERIES C.—FRENCH CANADIANS, CANADA, ALL LABORING PEOPLE.						
103	Boarding house, Montreal	8	100	77	573	3,475	7.5
104	Family, Montreal	3	141	177	723	5,190	8
105	Family, Quebec	8	132	100	513	3,575	5.6
106	Family, Quebec	8	136	127	641	4,365	6.8
107	Family, St. John	5	123	114	582	3,950	6.8
108	Family, St. John	10	82	91	431	2,950	7.8
109	Boarding house, Sorel	10	91	73	514	3,160	7.5
110	Boarding house, Sorel	12	86	86	473	3,090	7.8
111	Boarding house, Riviere du Loup	15	73	86	386	2,680	8
112	Family, St. Hyacinth	6	105	86	468	3,150	6.4
113	Family, Sherbrooke	4	150	168	536	4,375	6.1
114	Family, Richmond	8	86	91	477	3,155	7.9
115	Family, Richmond	5	104	100	522	3,495	7.2
	Average	102	108	106	526	3,585	7.1
	CONNECTICUT.[2]						
116	Family of chemist	5	118	103	430	3,205	5.6
	Family of college professor:						
117 a	Food purchased	4	129	183	467	4,145	6.8
117 b	Food eaten	4	128	177	466	4,080	6.8

[1] Collated by Massachusetts Labor Bureau.
[2] With the exception of Nos. 118 and 119 all Connecticut dietaries are in Middletown.

TABLE 37.—*Foreign and American dietaries*—Continued.

Reference No.	Dietaries.	Number of persons.	Nutrients.			Potential energy.	Nutritive ratio.
			Protien.	Fats.	Carbohydrates.		
	CONNECTICUT—continued.						
	Family of college professor, Storrs:						
	Winter dietary, January and February—		*Grams.*	*Grams.*	*Grams.*	*Calories*	*1:*
118 a	Food purchased	3	106	145	405	3,450	6.9
118 b	Food eaten	3	99	139	398	3,335	7.2
	Summer dietary, July—						
119 a	Food purchased	2	133	150	475	3,885	6.1
119 b	Food eaten	2	129	145	472	3,800	6.2
	Family of a retired merchant, son chemist:						
120 a	Food purchased	4	91	126	483	3,530	8.5
120 b	Food eaten	4	83	117	478	3,300	9
	Averages of Nos. 116 to 120; families of well-to-do professional men—						
	Food purchased	18	115	141	452	3,635	6.7
	Waste	18	4	5	3	75	......
	Food eaten	18	111	136	449	3,560	6.8
	M. club, college students; 3 dietaries of same club:						
	First dietary—						
121 a	Food purchased	24	161	204	680	5,345	7.1
121 b	Food eaten	24	138	184	622	4,825	7.5
	Second dietary—						
122 a	Food purchased	31	115	163	460	3 875	7.2
122 b	Food eaten	31	104	136	421	3,415	7
	Third dietary—						
123 a	Food purchased	27	113	180	376	3,680	6.9
123 b	Food eaten	27	92	141	346	3,110	7.2
	N. club, college students:						
124 a	Food purchased	3	141	160	503	4,130	6.1
124 b	Food eaten	3	140	158	503	4,105	6.2
	Students' club, divinity school:						
125 a	Food purchased	30	139	185	356	3,745	5.6
125 b	Food eaten	30	122	138	317	3,085	5.2
	Average of Nos. 121 to 125; boarding clubs of college and professional students—						
	Food purchased	115	134	178	475	4,155	6.6
	Waste	115	15	27	34	450	......
	Food eaten	115	119	151	441	3,705	6.6
	Average of Nos. 116 to 126, professional men and students—						
	Food purchased	133	125	160	464	3,905	6.6
	Waste	133	10	16	19	270	......
	Food eaten	133	115	141	445	3,635	6.7
	College football team:						
126 a	Food purchased	12	194	312	578	6,070	6.6
126 b	Food eaten	12	181	292	557	5,740	6.7
	Boarding house, well-paid machinists, etc.:						
127 a	Food purchased	21	126	188	426	4.010	6.8
127 b	Food eaten	21	103	152	402	3,485	7.3
	Family of a machinist:						
128 a	Food purchased	5	100	159	427	3,640	7.9
128 b	Food eaten	5	99	156	421	3,580	7.8
	Family of a blacksmith:						
129 a	Food purchased	4	103	176	408	3,730	7.8
129 b	Food eaten	4	100	171	401	3,640	7.9
	Family of a stone mason:						
	Winter dietary, December—						
130 a	Food purchased	3	107	153	391	3,470	6.9
130 b	Food eaten	3	104	148	375	3,350	6.9
	Spring dietary, April—						
131 a	Food purchased	3	125	145	366	3,365	5.6
131 b	Food eaten	3	119	137	348	3,190	5.5
	Family of a carpenter:						
132 a	Food purchased	3	125	152	498	3,970	6.7
132 b	Food eaten	3	114	135	475	3,670	6.9
	Family of a carpenter:						
	Winter dietary, December—						
133 a	Food purchased	3	107	161	408	3,610	7.2
133 b	Food eaten	3	100	149	388	3,390	7.3

TABLE 37.—*Foreign and American dietaries*—Continued.

Reference No.	Dietaries.	Number of persons.	Nutrients.			Potential energy.	Nutritive ratio.
			Protien.	Fats.	Carbohydrates.		
	CONNECTICUT—continued.						
	Family of a carpenter—Continued.						
	Spring dietary, May—		*Grams.*	*Grams.*	*Grams.*	*Calories*	*1:*
134 a	Food purchased	3	115	125	346	3, 055	5. 5
134 b	Food eaten	3	111	122	336	2, 965	5. 5
	Average of Nos. 127 to 134; wage-workers—						
	Food purchased	45	114	157	409	3, 605	6. 7
	Waste	45	8	11	16	200	......
	Food eaten	45	106	146	393	3, 405	6. 8
	PENNSYLVANIA.						
135	Miscellaneous families in poorest part of Philadelphia. Dietaries of 25 families:						
	Average	....	109	108	435	3, 235	6. 2
	Minimum (negro)	....	66	68	181	1, 630	5. 4
	Maximum (German)	....	202	206	608	5, 235	5. 3
	ILLINOIS.						
136	Miscellaneons families in poorest part of Chicago. Dietaries of 26 families:						
	Average	....	119	141	398	3, 425	6
	Minimum	....	86	100	213	2, 195	4. 6
	Maximum	....	168	204	626	4, 950	11. 3
137	United States army ration	....	120	161	454	3, 850	6. 8
138	United States navy ration	....	143	184	520	5, 000	7. 5

Nos. 1 to 3, Playfair, Chem. News, 1865 (XI), 221.
Nos. 4 and 5, Jürgensen, Ztschr. Biol., 1886, 489.
Nos. 6 to 8. Hultgren and Landergren, No. 6, Untersuchung über die Ernährung bei frei gewählter Kost, Hygiea, 1889; Festband No. 11, and Nos. 7 and 8, Untersuchung über die Ernährung schwedischer Arbeiter, Stockholm, 1891.
Nos. 9 to 11, Erismann, Arch. Hyg., 1889 (9), 23.
Nos. 12 and 13, Sarin, Ibid., 34.
No. 14, F. Hofmann, Ztschr. physiol. Chem., VI (1882), 372.
Nos. 15 to 23, Meinert, Armee und Volks-Ernährung, II, 189-221.
No. 24, Böhm, Meinert, II, 186.
No. 25, Forster, Untersuchung der Kost, 211.
No. 26, Von Rechenberg, Die Ernährung der Handweber, Leipsic, 1890.
Nos. 27 and 29, Schuster, Untersuchung der Kost, 165, 146.
Nos. 28 and 30, Richter, Ibid., 174.
Nos. 31 and 32, Forster, Ibid., 180, 191; Ztschr. Biol., 1873, 401.
No. 33, Pettenkofer and Voit, Ztschr. Biol., 1866 (2), 459.
Nos. 34 and 35, Beneke, Zur Ernährungslehre d. ges. Mensch., Cassel, 1878, 290.
Nos. 36 and 37, Ranke, Die Ernährung d. Menschen, 193, 230.
No. 38, Forster, Untersuchung der Kost, 213.
Nos. 39 and 40, Forster, Ztschr. Biol., 1873, 389, 390.
No. 41, Cramer, Ztschr. physiol. Chem., VI (1882), 346.
No. 42, Forster, Untersuchung der Kost, 208.
No. 43, Hoch, Dissertation. Rostock, 1888; Ref. Handbuch der Hygiene, Weyl., III, 1, 84.
No. 44, Voit. Ztschr. Biol., 1889, 232.
No. 45, Voit, Untersuchung der Kost, 28.
Nos. 46, 49 and 51, report of Royal Bavarian special commission, see Meinert, II, 225.
Nos. 47 and 48, Forster, Ztschr. Biol., 1873, 387, 388.
No. 50, Steinheil, Ztschr. Biol., 1877, 415.
No. 52, Ranke, Ibid., 130.
No. 53, Prausnitz, Arch. Hyg., 15 (1892), 387.
No. 54, Hößler, Der Isarwinkel, Munich, 1891, 63-72.
No. 55, Hößler, Ibid., and Liebig, Meinert, II, 224; Sitzungsber. d. bayr. Acad., II, 463, 1869; Reden u. Abhandl., 121.
Nos. 56 to 59, Meinert, Armee und Volks-Ernährung, I, 276-287; and König, third edition, I, 156.
No. 60, Ohlmüller, Ztschr. Biol., 1884, 393
Nos. 61 and 62, Manfredi, Arch. Hyg., 17 (1893), 552.
Nos. 63 to 65, Albertoni and Novi, Arch. Physiol., 1894, 213.
No. 66, Moleschott, Razione del Soldato Italiano, Rome, 1883.
No. 67, Beaunis, Recherches Expér., Paris, 1884, 4; Ref. Handbuch der Hygiene, Weyl., III, 1, 84.

Nos. 68 to 71, Forster, Ztschr. Biol., 9 (1873), 381. Camerer, Ibid., 1880, 24; 1882, 220, 1884, 556; 1888, 141; 1893, 227, 398. Uffelmann, Ibid., 1882, 584. Hasse, Ibid., 1882, 553.
No. 72, Voit, Untersuchung der Kost, 125.
No. 73, Schröder, Arch. Hyg., 4 (1866), 39.
No. 74, Prausnitz, Ibid., 15 (1892), 387.
Nos. 75, 76, and 80, Eijkman, Ztschr. Biol., 1889, 106.
Nos. 77, 81, and 82. Tawara, Ibid., 1889, 107; Arch. Hyg., 1888, 102.
No. 78, Kellner and Mori, Ztschr. Biol., 1889, 102.
No. 79, Scheube, Arch. Hyg., 1883, 352.
No. 83, Atwater and associates.
Nos. 84 to 115, Atwater and assistants. Food consumption; Seventeenth Annual Report Massachusetts Bureau of Statistics of Labor. Report of Storrs (Conn.) Experiment Station for 1891, 106.
Nos. 116 to 134, and 137 and 138, Atwater and associates, Reports of Storrs (Conn.) Experiment Station, 1891, 106; 1892, 135; 1893, 174.
Nos. 135 and 136, Shapleigh. A Study of Dietarics; Partial Report of Dutton Fellow, College Settlements Association, 1892-93.

COMMENTS UPON DIETARIES SUMMARIZED IN TABLE 37.

In the belief that the figures of Table 37 will be best understood by statement of some of the more important details, an attempt to recapitulate the latter is made in the pages which follow.

English dietaries.—Nos. 1, 2, and 3 are those estimated by Sir Lyon Playfair from the actual weights of the food consumed by individuals; Playfair speaks of the engineers (soldiers) of No. 11 as "laborers in time of peace actively occupied either in the construction of field works, or pursuing their avocations as artisans," and regards the dietary as "the most complete evidence we possess [at the time, 1864–1865] of the requirements of food for laboring men doing a fair but not excessive amount of work in 24 hours; when these soldiers are at light labor they are found to take less." The sailors of No. 1 and weavers of No. 2 were prisoners in the Wakefield Gaol. The details of food consumption were very carefully observed, and are published by Dr. E. Smith.[1]

Danish and Swedish dietaries.—Nos. 4 and 5 are reported by Jürgesen. No. 4 is that of a physician in Stockholm, 37 years old. The food in this case, as in No. 5, consisted of milk, meat, fish, bread, cheese, butter, and beer. The investigation was made in January and February. No. 5 is that of the wife of the physician just named, 35 years old, principal of a girls' school; time of investigation, May. The food actually eaten was weighed and its composition estimated from the compilations of König and Almen.

Nos. 6, 7, and 8 are reported by Hultgren and Landgren. The food materials were carefully weighed as used; the composition was estimated from analyses published by König and Almen. For substances of which no reliable figures for composition were available, as pastry, etc., special analyses were made.

No. 6 represents the average of 5 dietaries of as many medical students in Stockholm, from 22 to 27 years of age, busily engaged in study and laboratory work. The examinations were made in the winter. The experimental periods ranged from 8 to 16 days; the protein per person per day from 103 to 163 grams, and the energy from 2,800 to 3,375 calories.

No. 7 represents the average of 6 dietaries of 4 healthy men, a metal worker, a blacksmith, a carpenter, and a farm laborer, from 28 to 46 years of age, at moderately hard work and earning good wages. The food included meat, fish, milk, etc., and alcohol in beer or stronger spirits. The alcohol averaged 22 grams per man per day, and is estimated as equivalent in fuel value to 37.8 grams of carbohydrates. The protein ranged from 105 to 166 grams and the energy from 2,895 to 3,510 calories per man per day. The diet of this group of men is regarded by the authors as characteristic in respect to nutrients and energy for Swedish laboring men at moderate work (mittlere Arbeit).

No. 8 represents the average of 5 dietaries of 5 vigorous men, a stone mason, a bricklayer, a carpenter, a wood sawyer, and a farm laborer, from 31 to 54 years of

[1] Phil. Trans., vol. 151, 747.

age, at very hard work, and earning good wages. The food was similar to that of No. 7. The alcohol averaged 24.2 grams per man per day. It was estimated as equivalent to 41.3 grams of carbohydrates, and included in the estimates of carbohydrates and energy as in No. 7. The protein ranged from 128 to 246 grams, and the energy from 3,655 to 5,580 calories per man per day. The authors regard the diet of this group of men as fairly representative of that of Swedish workingmen at severe work.

Russian dietaries.—Nos. 9, 10, and 11 are dietaries of factory operatives in the neighborhood of Moscow, reported by Erismann. It is the general custom of Russian factory operatives, such as are represented in these studies, to board in clubs of their own organization. The purchasing of food, etc., is delegated to a committee, which also keeps a careful account of the number of meals eaten by each member of the club. These accounts are kept with the greatest care, as all the bills are paid by the owners of the factory, and a sum proportional to the number of meals eaten by each member of the club deducted from his wages. The statistics upon which these dietaries are based are derived from these accounts, and represent the quantities of food purchased during periods of from two to three months in length. Their food consists principally of black bread, buckwheat grits, and sauerkraut, and during those parts of the year when fasts are not observed, meat and animal fats. During the fasts, which cover a considerable part of the year, vegetable oils are used to a considerable extent. The composition of the food was estimated principally from previous analyses by Russian observers, otherwise according to König's compilations.

Nos. 12 and 13 are dietaries of peasants in the same locality, by Sarin. Each study occupied one week The food actually eaten by each person at each meal was carefully weighed; the composition was estimated.

German dietaries.—No. 14 is the dietary of the family of an intelligent laboring man, who had not been able to obtain work for several months, and was almost destitute at the time of the inquiry. The study was made by F. Hofmann.

Nos. 15 to 23 are dietaries of laboring people of the poorer classes, living, No. 23 near, and the rest in the city of Leipsic, Saxony. The inquiry was conducted in part, at least, and presumably the whole, in 1878-1880, by Dr. Meinert, with the purpose of learning something of the conditions of life, and especially the food of people of small incomes in Germany. It was evidently conducted with no little care and thoroughness. For each dietary the food was weighed during a period of five or seven days, with such detail as to show the amounts of each food material per meal. Allowance was made for the waste in one case, at any rate, as stated in the report and presumably for all, so that the quantities given appear to be those actually consumed. No statement is made of analyses especially executed for the investigation, and it is to be inferred that the composition of the food was estimated from standard analyses. The quantities of nutrients and costs of food are estimated per adult per day. In the cases where there were children the equivalent number of adults was assumed for the calculations. The amounts of food of different kinds eaten per meal and per day, the amounts and costs of the several kinds of food, and of the whole food of each dietary, the incomes of the people, and their ways of living, were observed and reported. While the quantities of nutrients imply straitened circumstances in most of the cases, the details in some are decidedly pathetic. The space here will allow only a brief summary of some of the more important points.

No. 15, girl in cigar factory, 25 years old, earning an average of 5 marks per week, and boarding with her mother, who earned 3 marks per week by washing. Their food consisted mainly of bread, potatoes, a little rice, occasionally vegetables, meat 250 grams for 2 persons twice a week, or half a pound a week for each person, and coffee. In lack of fuel for heating the coffee, hot water from a neighboring factory was used. The description says "it is not strange that with food so poor in protein both mother and daughter are frail and sickly, nor that, as the girl told me in tears, they often went supperless to bed. * * * The girl is the mother of a child 6 months old. * * * For board and lodging of herself and child she pays 3½ marks per week, and thus has left from her wages for clothing and other expenses for herself and child 1½ marks per week, a sum which, while it does not justify, explains her increasing it by other means." The amounts and composition of the food are computed for the young woman and her mother, so that the figures represent the average for the two.

No. 16, girl working in a printing office. She was 29 years old, her wages averaged 5 marks per week, she lived with another girl and a child of the latter in a room 10 feet square. In the three winter months of 1879–80, she had only bread and butter and coffee, and for dinner daily from 1¼ to 2½ cents' worth of sausage, and even on this fare used up $5.50 of previous savings. The amount, composition, and cost of the food were estimated for the two girls together, so that the figures represent the average. "What wonder," says Dr. Meinert, "that young women, whose moral energy and intellectual powers are reduced by such a starvation diet, are unable to withstand temptation."

No. 17, widow, with five children, aged 5½, 8, 10, 12, and 14 years, respectively. She supported herself and children by plaiting straw and taking boarders. The food consisted of bread, beef fat, potatoes, vegetables, and occasionally meat. Family estimated equivalent to 4 adults.

Nos. 18 and 20. No. 18, girl, 24 years old; in paper factory; wages, 9 marks; sister, 21 years; seamstress; wages, 8 marks per week. No. 20, brother of preceding, 28 years old; printer; wages, 18 marks per week. These persons lived with their invalid mother in a fifth-story apartment, consisting of a living room, a sleeping room, and kitchen. Dr. Meinert says that with their earnings these people might have had much better food. "The appearance of the two girls, however, explained the scantiness of their diet. Their clothing, especially that of the seamstress, was much finer than their position and income warranted, and formed a sadly striking contrast to the appearance of the mother and the pitiable dwelling in which the people lived. The food included meat daily. Account was taken of the food of the man and of that of the women separately, the man having more than the women.

Nos. 19 and 21. No. 21, family of cabinetmaker, consisting of father and 6 children, aged ⅓, 3, 7, 9, 11, and 14 years, and father's sister (No. 19), a girl of 18, working in a bookbindery. The cabinetmaker earned 18 marks and the sister 9 marks per week. The 9 persons lived in an apartment consisting of 2 rooms and a closet holding a cook stove. The sister paid for lodging, breakfast, and dinner 3½ marks (84 cents) per week and purchased bread and butter extra. The wife and children were undersized, pale, and weakly.

No. 22, family of druggist's clerk, consisting of father, mother, and 2 daughters, aged 16 and 12. They are spoken of as thoroughly good people. The annual income was 1,180 marks. Their food was better than the preceding. It included more variety; vegetables, butter, and meat were eaten daily; indeed the family had meat, eggs, or cheese once and often twice a day. As the man's labor was light, Dr. Meinert regards his diet as sufficient for his needs.

No. 23, family of farm laborer near Leipsic, consisting of father, mother, and 4 children, from 5 to 14 years old. The 6 persons are accounted as equivalent to 4 adults. The food consisted mainly of bread, beef fat, and potatoes, with a little milk, and for one meal in the week, Sunday dinner, meat.

The final statistics are summarized in the following table:

Food materials and costs per adult per day.

Reference No.	Dietaries.	Food materials.	Nutrients. Protein. Total.	Nutrients. Protein. Digestible.	Nutrients. Fats.	Nutrients. Carbohydrates.	Cost.	
		Grams.	*Grams.*	*Grams.*	*Grams.*	*Grams.*	*Pfennig.*	*Cents.*
15	Girl in cigar factory	839	52	42	53	301	34	8.2
16	Girl in printing office	1,202	65	54	39	303	43	10.3
17	Woman at straw plaiting	1,504	72	56	56	440	40	9.6
18	Girl in paper factory and seamstress	812	56	47	51	229	36	8.6
19	Girl in bookbindery	1,062	61	49	41	347	36	8.6
20	Printer	1,195	87	73	64	366	48	11.5
21	Cabinetmaker	1,271	77	62	57	466	46	11
22	Druggist's clerk	1,129	71	59	69	351	47	11.3
23	Farm laborer	1,394	80	61	37	504	30	7.2

No. 26, hand weavers in Zittau, Saxony. For detailed description see pp. 164–177.

No. 27 to 30 are dietaries of prisoners in Munich reported by Schuster, and in Brandenburg, Prussia, by Richter. No. 27 is that of a house of detention and No. 28 of a house of correction in Munich; in the latter the terms of service were 3 years or longer with labor. No. 29 is the ration served to prisoners without work, and No. 30 that of inmates of a house of correction, with labor, in Brandenburg. In all the prison dietaries the food was mostly vegetable, and in Munich it was notably insufficient in quantity. In Nos. 27 and 28 the rations were carefully weighed and analyzed. In Nos. 29 and 30 the rations were weighed as prepared for cooking, but no analyses were made.

Nos. 31 and 32 are reported by Forster, and are the dietaries of inmates of infirmaries in Munich for aged and infirm persons in part or wholly incapable of self support. In Nos. 31 and 32 and in all of Forster's dietaries all the food was carefully weighed and samples taken for a more or less complete analysis. When actual analyses were not made the composition was estimated from analyses by Voit and others.

No. 33 is the average of three tests in the respiration apparatus made by Pettenkofer and Voit. The subject was a watchmaker in Munich, 28 years of age, and weighing 70 kilograms (152 pounds). During the experiments he did no muscular work and lived upon an ordinary mixed diet. The quantities of nutrients are those actually metabolized.

Nos. 34 and 35 represent observations by Professor Benecke, of the University of Marburg, upon himself. He was in decidedly active mental and light physical work, had about 3 hours' exercise in the laboratory and 8 hours' sleep. With a mixed diet of meat, milk, bread, etc., containing the nutrients of No. 34, his weight decreased 451 grams (1 pound) during an experiment of 14 days. From determinations of nitrogen in food and excreta, he estimated that the dietary of No. 35 would have sufficed for his needs.

Nos. 36 and 37 represent the dietaries of Professor Ranke, of the University of Munich, in experiments upon himself, in which the total income and outgo were determined with a respiration apparatus. With very little physical exercise these quantities just sufficed him for the maintenance of his body without gain or loss.

Nos. 38 to 40 were reported by Forster. The Munich lawyer, No. 38, was in moderate circumstances, but able to have all the food that health demanded. The quantities here given are one-third of those consumed by a family of three adults. One-fourth of the carbohydrates of the food of the physician was taken in beer.

Nos. 34 to 40 were the dietaries of professional men in comfortable circumstances. Their food differed from that of the preceding and most of the succeeding dietaries in that it had much more of meat and other animal foods, and was hence more completely digestible; and that it had more variety, which made it more palatable.

No. 41. In contrast with these six dietaries of professional men, Nos. 35 to 40, which were composed largely of animal food, may be cited that of an official of high rank in the German civil service, reported by Cramer. The official was 64 years of age, and had been a vegetarian for 11 years. He was in good health and possessed of a great deal of physical endurance. The character of the food in this dietary differs considerably from that in No. 44, as the official was not a so-called strict vegetarian, but used milk, eggs, and warm food. The food was weighed and its composition determined by analysis. Digestibility was also determined.

No. 42 was reported by Forster. This mechanic was connected with a laboratory, was in very comfortable circumstances, had considerable animal food, and did very light muscular work.

No. 43 is the dietary of a shoemaker in Rostock, observed by Hoch.

No. 44 is the dietary of an upholsterer, 28 years of age, in Munich. He was a strict vegetarian, and had lived for 3 years upon an exclusively vegetable diet, containing no warm food, and consisting of bread, fruit, and oil. The man was normally developed, appeared to be well fed, and weighed 57 kilograms (125 pounds). This dietary is reported by Voit, and the food was weighed and analyzed. The digestibility of the food was also determined.

No. 45 represents the average result of examinations by Voit of the food of 3 well-paid mechanics in Munich. It consisted of meat, bread, other vegetable food, and beer. The quantities and composition of the food were estimated.

No. 46 is the average of 2 dietaries of locksmiths, 7 of carpenters, and 3 of coopers in Bavaria, given in a royal Bavarian commission report.

Nos. 47 and 48 were also reported by Forster.

No. 49 is the average of 5 dietaries of Bavarian farm laborers reported by the same commission referred to in No. 46.

No. 50 is the average result of examinations by Steinheil of the food of miners near Ems, Prussia, at severe work. The food included meat, bread, potatoes, beans, and other vegetables in considerable variety, and butter. It is probable that actual weighings of the food were made; its composition was estimated presumably from Voit's compilations.

No. 51 is the average of 6 dietaries of Bavarian brewery laborers, reported by the Bavarian commission above mentioned. Their work is very severe, and they are noted for consuming large amounts of food, of which beer furnishes a considerable proportion of the carbohydrates.

The figures for the individual dietaries of which Nos. 46, 49, and 51 express the averages, are as follows: The protein in the mechanics' dietaries varies from 47 to 183 grams, and the energy from 1,690 to 5,285 calories. I have not the original data at hand, but the source would seem to vouch for their reliability. The range is very wide, but I am inclined to think not wider than might be found among people of similar occupation in ordinary communities.

Dietaries of workingmen in Bavaria.

	Nutrients.			
	Protein.	Fats.	Carbohydrates.	Potential energy.
	Grams.	*Grams.*	*Grams.*	*Calories.*
Locksmith	109	41	407	2,495
Do	78	12	331	1,790
Carpenter	149	25	974	4,835
Do	163	30	1,058	5,285
Do	173	59	513	3,360
Do	183	52	761	4,355
Do	65	22	228	1,405
Do	99	52	585	3,290
Cooper	124	22	651	3,380
Do	47	7	350	1,690
Do	147	50	409	2,745
Average of 11 dietaries of mechanics	122	34	570	3,150

Dietaries of workingmen in Bavaria—Continued.

	Nutrients.			
	Protein.	Fats.	Carbohydrates.	Potential energy.
	Grams.	*Grams.*	*Grams.*	*Calories.*
Farm laborer	131	61	609	3,600
Do	123	61	557	3,355
Do	189	63	683	4,160
Do	95	33	304	1,940
Do	147	59	559	3,445
Average of 5 dietaries of farm laborers	137	55	542	3,300
Brewery laborer	197	63	897	5,070
Do	122	34	674	3,580
Do	85	32	728	3,630
Do	116	61	566	3,365
Do	223	113	909	5,690
Average of 5 dietaries of brewery laborers	149	61	755	4,270

No. 52 is the dietary reported by Ranke of a large number of Italian brickmakers working near Munich during the summer. Some three hundred were in charge of a superintendent, who provided them with food which consisted of maize meal and cheese, with occasionally a small amount of brandy. The labor was decidedly severe. The amounts of food materials were based upon statistics furnished by the superintendent. The composition was estimated by Ranke, but his assumed composition of maize evidently does not represent the maize of this region. The amounts of nutrients given in the table are recalculated, using for the composition of maize the averages of analyses of maize grown in southern and southwestern Europe.[1]

No. 53 is the dietary of machinists, etc., employed in the Krupp gun works at Essen, reported by Prausnitz. It has been for many years the endeavor of this firm to do all in its power to improve the food and quarters of its employees. The boarding houses (menage) were established in the interest of the unmarried workmen and those whose families lived at a distance, the purpose being to furnish them with suitable board and comfortable lodging. They were opened in 1856 with 200 members; this number gradually increased with the extension of the works, until at the time of the investigation in 1890 it had reached about 800. Since 1884 all unmarried workingmen, except those receiving high wages for special work and those living with their relatives, have been obliged to live in the company's boarding houses, a regulation which has been advantageous to both boarding houses and boarders. The price for dinner and supper together is 80 pfennigs (19 cents) per day. This does not include bread, but each man receives one-eight kilogram of coffee and one-fourth kilogram of butter per week. At dinner and supper the meat is served in individual portions; the men are allowed to take as much as they choose of all other dishes.

The average quantities of nutrients furnished in the dinner and supper are 115 grams protein, 81 grams fats, and 480 grams carbohydrates per day. In addition to this an allowance of 400 grams bread and 36 grams butter is made for the breakfast, which brings the total nutrients up to the amounts stated in the table. Prausnitz regards these figures to be the minimum estimate, as the daily amount of bread is placed very low, and milk, sugar, and beer are not included; this would fully balance all food wasted in preparation and at meals. The quantities of food materials are estimated on the basis of the bills of fare and their composition from König's averages.

No. 54. This represents the average of 6 dietaries examined by Höfler of farm laborers in the district of Tölz, in upper Bavaria. The dietaries are included in a very interesting description of the region and the people. The author says, in speaking of the peasants of the district and their very unusually ample nourishment:

[1] König, Chem. d. mensch. Nahr. u. Genuss-Mittel, third edition, I, 559.

"The influence of their food upon their bodily development is very noticeable. The people are tall, extremely strong, and sinewy; they are not fat, but their organic albuminoids (muscular as distinguished from fatty tissue) are well developed. * * * Really 'fat' men, that is to say, men with large deposits of fat in the body, are rare among the peasantry of this region. They have large and muscular frames and are among the best built men in the army enlistment rolls. * * * Aside from their severe work, a still further reason for the peculiarity of their food (large total amount and especially large quantities of fat) is found in the climate. * * * The low temperature and the constant wind make the large amounts of fat a necessity. * * * People who have to endure such ranges of temperature as prevail here, without especial protection by clothing, and not suffer in health, must be enabled to regulate the bodily warmth by large supplies of easily combustible fats and carbohydrates in the food."

The food consists of bread, flour, potatoes, more or less garden vegetables, milk, cheese, grease, and butter. The protein, outside that of the bread and flour, comes mainly from milk, cheese, and curd, the fats from grease and butter which are consumed in large quantities, and the carbohydrates from wheat, barley, and potatoes. Sugar is little used because of its cost. The description does not state how the dietaries were estimated, but only gives the quantities of nutrients as stated beyond.

No. 55 is the average of 3 dietaries of lumbermen—1 from Tölz, as reported by Höfler, and 2 from other parts of upper Bavaria, as reported by Liebig. These are in size and amounts of fat and of energy entirely exceptional among the European dietaries of which I have been able to find record, and rank even with the larger, though they are far from approaching the largest, of the American dietaries given in the previous pages. Of the peasantry in the region where the 2 dietaries of lumbermen by Liebig were taken, Professor Ranke says:

"In contrast to the conditions prevalent elsewhere in Germany, the food of the country people in the Bavarian highlands and mountains is very ample, and although potatoes have not become the principal food material here, the food is chiefly vegetable. The famous peasant of the Bavarian mountain region, the genuine 'Haberfeldtreiber,' as he proudly calls himself, eats meat, in accordance with time-honored usage, only on the four great holidays of the year. He lives upon 'Schmalzkost,' that is to say, simple preparations of flour with which large quantities of fat are incorporated. To these he adds such materials as sauerkraut and dried apples or peas. The food of these powerful peasants is so generous as to explain their herculean development of muscle, their enviable vigor, and their consciousness of strength, which often leads to excess. * * * It is likewise a current observation by people in the Tegernsee Mountains that the lumbermen can do the more work the larger their appetites are."

In comparing the dietaries of the Bavarian lumbermen, as reported by Liebig, with other European dietaries, I have sometimes, in my own mind, questioned whether the figures were entirely reliable as exponents of the eating habits of the class to which the men belonged. In matters of larger importance some of Liebig's doctrines have been doubted, only to be confirmed by the results of later inquiry. In this minor though interesting matter of detail the teaching of the great master is substantiated by the figures which Höfler cites. It would certainly be a valuable contribution to the science of nutrition if a number of these peasant dietaries could be thoroughly studied.

The table herewith summarizes the data of the dietaries just mentioned.

Dietaries of men in Bavarian highlands.

Occupation and locality.	Nutrients.			Potential energy.
	Protein.	Fats.	Carbohydrates.	
FARM LABORERS.	*Grams.*	*Grams.*	*Grams.*	*Calories.*
Jachenau, Tölz, average of 3 estimates (Hofler)	161	195	467	4,390
Gaissach, Tölz, average of 2 estimates (Höfler)	166	201	734	5,560
Kirchbichl, Tölz, one measurement (Höfler)	85	210	438	4,095
Average of 3 dietaries of farm laborers (Höfler)	137	202	546	4,680
LUMBERMEN				
Jachenau, Tölz (Höfler)	144	358	606	6,405
Reichenhall Mountains (Liebig)	112	309	691	6,165
Oberaudorfer Mountains (Liebig)	135	208	876	6,080
Average of 3 dietaries of lumbermen	130	292	724	6,215

No. 56 to 59. Army rations, like other dietaries, are decidedly variable. The figures here are borrowed from very extensive compilations by Meinert. No. 56 is the average of 4 estimates by Voit, Artmann, Hildersheim, and Meinert, respectively. No. 57 is the average of 3 estimates by Artmann, Hildersheim, and Meinert. No. 58 is an extraordinary ration for soldiers in active service in the field.

German army management, from Frederick the Great to William and Moltke, has recognized the fundamental principle that soldiers to fight well must be well fed. No. 59 is therefore especially interesting as a ration ordered shortly after the outbreak of the Franco-German war, at the time of the terrible marching and fighting which brought the victories at Worth, Metz, and Sedan.

Austria.—No. 60 represents the average daily food consumption of 15 farm laborers in Central Transylvania as observed by Ohlmüller. The men were engaged in the severe work of harvesting from 4 o'clock in the morning until evening; their food consisted exclusively of maize meal and beans, and their only drink was water. The amounts of nutrients were estimated by Ohlmüller, who used the same composition for maize which Ranke used in No. 52 (Italian brickmakers). The nutrients are, therefore, recalculated, the same composition being assumed for maize as in No. 52.

Italian dietaries.—Nos. 61 and 62 are the dietaries of people of the poorer class in Naples, as observed by Manfredi and described above. No. 61 is the average of 5 dietaries of men, 2 shoemakers, a carpenter, a mason, and 1 without employment. No. 62 is the average of 3 dietaries of women (servants). Nos. 61 and 62 are described in detail on pages 173–177.

Nos. 63 to 65 are the average dietaries of a peasant 39 years of age, his wife, and son 14 years of age—a shoemaker—who lived in the neighborhood of Ferrara. They were studied by Albertoni and Novi. The family was in good health but very poor; their combined yearly earnings amounted to only $97, of which $81 was expended for food. Their food in winter was mainly maize and chestnut meal, macaroni, beans, fish, and lard. In summer the maize meal was in part replaced by bread. The food was accurately weighed and analyzed, as were also the urine and feces. Each of the 3 dietaries represents 2 periods of observation of 3 days each, one in March and the other in August.

No. 66 is the average of 4 Italian army rations, 2 for infantry and 2 for cavalry, in time of peace, as calculated by Moleschott.

France.—No. 67 was reported by Beaunis.

Children (Germany, Russia, and Switzerland).—Nos. 68 to 71 are averages of individual dietaries of children, with one or two exceptions in well-to-do private families. Of the 38 dietaries included in these averages 2 were observed by Forster, 3 by Uffelman, 10 by Hasse, and 23 by Camerer, all physicians. Forster's dietaries were those of 2 infants. One was in a laborer's family; it was 7 weeks old and was fed on a porridge made of milk, flour, and sugar. The other belonged to a well-to-do family; it was between 4 and 5 months old and was fed condensed milk. Both these children probably lived in Munich. Uffelman observed the dietaries of his own boys, from 2 to 15 years of age, in Rostock; only those of the three younger ones are included in these averages. Hasse made two series of studies of the dietaries of 4 girls, daughters of a well-to-do family in St. Petersburg; their ages ranged between 2 and 11 years. They were very well developed and their dietary consisted largely of animal food. The food was carefully weighed and its composition in parts determined by direct analysis and in part estimated from König's compilations. Hasse also observed the dietaries of two girls, aged 2 and 4 years, in a well-to-do family of Russian descent in Zurich, Switzerland. The food was essentially the same as that of the children in St. Petersburg and the method of the investigation the same. In all of Hasse's dietaries the period of observation ranged from 3 to 6 days.

Camerer made all told 7 series of dietaries of his own children, 4 girls and 1 boy, from 1878 to 1892. In 1878 the ages of the girls were 2, 3½, 9, and 11 years, and that of the boy 5½ years. In 1892 those of the girls were 15, 16, 22, and 24 years and that of the boy 18 years.

The general plan of these investigations consisted in observing the food of each child for 6 periods (in a few cases a less number), of 4 days each during each year. The 6 dietaries thus obtained were averaged together for the dietary of the child for the year, and these average (annual) dietaries are the ones included in the averages in the table.

Except in one of the 7 annual series weighings were made of the food actually eaten and its composition in part determined by analysis and in part estimated from König's compilations. In addition to these data the weights of the children were observed from day to day, the urine was collected and the urea determined and the feces collected, weighed, and analyzed. These studies were made in Riedlingen and Urach, in Würtemberg.

During the period between 1878 and 1892, 31 of these average annual dietaries were made, representing approximately 37 observations for each child or 186 for the five children. Only those dietaries of the children up to 14 years of age are here included. Of these there were 23 annual dietaries representing approximately (6×23) 138 observations of 4 days each. It should be added that the children were healthy and their growth was normal, although the girls who had attained the ages of 15 to 20 before the end of the study were somewhat smaller than the average.

These are by far the most extensive and valuable observations upon the food consumption of children which have been reported.

No. 68 is the average of the 2 dietaries reported by Forster, referred to above, and 1 by Camerer, that of a child not in his own family. This girl was 1 year and 2 months old and weighed 23 pounds. Her food was cows' milk, toast, gruel, beef tea, etc. The food was weighed and its composition in part determined by analysis and in part estimated from König's compilations. The average age of these 3 children was 8 months and their average weight 16 pounds.

No. 69 is the average of 14 dietaries, 9 of which are those of girls and 3 of boys from 2 to 6 years of age. Of the girls' dietaries 6 were reported by Hasse and 5 by Camerer. Two of the 3 boys were Uffelman's and the other Camerer's. The average age of these 14 children was 3 years and 10 months and their average weight 33 pounds.

No. 70 is the average of 8 dietaries, 6 of girls and 2 of boys from 6 to 10 years of age. One of the girls' dietaries was reported by Hasse. All the other dietaries are those of Camerer's children. The average age of the 8 children was 8 years and 5 months and their average weight 53 pounds.

No. 71 is the average of 13 dietaries, 10 of girls and 3 of boys between the ages of 10 and 14. Eight of the girls' and 2 of the boys' dietaries were observed by Camerer, those of the other two girls by Hasse, and that of the boy by Uffelman. The average age of the 13 children was 11 years and 8 months and their average weight 70 pounds.

No. 72 is the dietary of children 6 to 15 years old, inmates of an orphan asylum in Munich reported by Voit, who pronounced the food satisfactory both as regards quality and quantity. The food was weighed as prepared for cooking, but no mention is made of actual analyses.

No. 73 is a dietary studied at a home for children in Gehlsdorf, near Rostock, containing 38 boys between 8 and 15 years of age. They were healthy and normally developed, and occupied in study and light work. The study was made by Schröder. The food was weighed as prepared for cooking and its composition was estimated. The rules of the institution did not permit waste of food at the table.

No. 74 is the dietary of girls from 14 to 19 years of age (average, 15½ years) in the Krupp training school for girls. This school represents one phase of the effort which has been made by the Messrs. Krupp to elevate the condition of their employees. The city of Essen, which has nearly 70,000 inhabitants, depends mainly upon the Krupp cast steel works, which employ some 10,000 workmen in the shops and 8,000 outside. While this establishment is of supreme interest to the mechanical engineer, the efforts of the Messrs. Krupp to promote the welfare of their employees are scarcely less so to the student of sociology.

The girls' training school has for its object not only the general education but instruction in those things which pertain to the care of the household and the duties of the wife and mother. The course in these latter branches lasts for 3 months, and includes purchasing of food, cooking, preserving and care of provisions, management of the kitchen garden, washing, ironing, mangling, knitting, darning, and all kinds of housework. In instruction in food economy the pupils are taught what quantities of food materials are required for a given number of persons and how much should be prepared for each meal. Each pupil is expected to provide for 10 others for a number of days. She must weigh out the necessary food materials and prepare and cook them. She must keep an exact account of what is used for each meal and the cost and quantities of the several materials employed. At the end of the course each girl takes her own account book, which makes a very valuable guide for her later in life, in that it tells her how a nutritious and at the same time varied diet may be furnished at a comparatively small expenditure. This dietary study was made in connection with No. 53, by Prausnitz. The quantities of food materials were taken from the girls' account books mentioned. The composition was estimated from König's compilations.

Japanese dietaries.—According to Mori 3 general classes of dietaries are common among the Japanese, namely, (1) that of the rural population of the interior. This is almost exclusively vegetable; fish is eaten but once or twice a month and meat but once or twice a year; (2) that of the population of the coast, who eat fish in considerable quantities, and (3) that of the city population and of well-to-do families, who eat both meat and fish to a considerable extent.

Rice is the principal article of vegetable food, but, in addition to this, barley, wheat, various kinds of millet, and buckwheat are eaten in considerable quantities. Tubers and roots, such as turnips and radishes, are staple articles of food, and pumpkins, cucumbers, etc., are much used. The legumes are little eaten in their natural state, but form the basis of a number of prepared foods and relishes, such as miso, tofu, and shoyu, all of which are made from the soja bean. The miso is prepared from cooked beans, which are rubbed to a thick paste and fermented with the ferment used in the preparation of the rice wine. Tofu, or bean cheese, is essentially the legumin of the soja bean, which is first extracted with water and then precipitated by the addition of the mother liquor (magnesium chlorid), obtained from the evaporation of sea water in the manufacture of salt. The cheese is eaten fresh. The shoyu sauce is prepared from a mixture of cooked and pulverized soja beans, roasted and pulverized wheat, wheat flour, salt, and water. The mixture is fermented with the above-mentioned rice ferment for 1½ to 5 years in casks. This sauce is used very largely by all classes.

Nos. 75 and 76 are dietaries of prisoners in Tokyo, observed by Eijkman. The food was exclusively vegetable, and consisted mainly of a mixture of one part of rice and one and one-half parts of barley cooked together. All of Eijkman's dietaries, Nos. 75, 76, and 80, were published in Japanese. Kellner and Mori and Nakahena in referring to them state that the food was in all cases weighed and its moisture content determined.

No. 77 is that of the employees in a large retail store in Tokyo, reported by Tawara. The period of observation covered only 3 days, but during this time no animal food was eaten; the food was essentially rice and salted radish. Tawara's dietaries, Nos. 77, 81, and 82, are also published in Japanese, but are cited by the writers above referred to, who state that the food was weighed and its composition estimated upon the basis of numerous analyses of Japanese food materials by Tawara and others.

No. 78 is the dietary of Mori, an assistant at the University of Tokyo. It was observed by himself, and he regards it as representative of the food of people in well-to-do circumstances and students in the higher institutions. It is materially modified by European influences (more meat and more protein), but on account of its expense has gained little footing among the majority of the people. The food was beef, rice, potatoes and other vegetables, and milk. The test was continued for 6 consecutive days, during the last 3 of which the food materials and also the urine and feces were analyzed. The amounts given are, however, those actually eaten.

No. 79 gives the average of the dietaries of two students in the University of Kioto, as observed by Scheube. Meat and fish were eaten daily, and eggs, bean cheese (tofu), and green vegetables were also used. Each test was continued for 5 days. The food was weighed and its composition estimated.

No. 80 is the dietary of a military school in Tokyo, reported by Eijkman. The observations cover a period of several nonconsecutive days. Meat was served daily.

No. 81 is that of the inmates of a Government school in Tokyo, from 17 to 25 (average, 20) years of age. The cost of the food was $4 per month per pupil. Meat was served once a day, and fish was eaten in considerable quantities. The period of observation covered 10 consecutive days.

No. 82 is the dietary of the pupils in a private school, also in Tokyo. Their ages ranged from 11 to 21 years. Meat was served only twice a week in this school, but fish was eaten regularly. The price of the food was about $2.25 per month per pupil. Their food consumption was observed during one week. Nos. 81 and 82 are reported by Tawara. In the opinion of Kellner and Mori the quantities of food served to these pupils, of whom a considerable number were children, were somewhat excessive, and doubtless there were uneaten residues, especially of rice, which were not taken into account in estimating the amounts of nutrients. This would appear to be the case especially in No. 82.

Java.—No. 83. In connection with the study of the food of some of the interesting people from the less-known countries at the World's Fair, made under the auspices of the World's Columbian Commission,[1] an opportunity was furnished to observe with exactness the food consumption of several inmates of the Java village. One of the houses was set apart for the purpose, and 2 families, including 2 women

[1]A report of these investigations, which were conducted under the direction of the author, will, it is expected, appear with other reports of the World's Fair.

and 3 men, were assigned to live in it for a period of 10 days. They were enjoined to eat nothing except what was served in the house, and the cooks were charged to provide the fullest opportunity for collecting the data desired.

The men were engaged to a certain extent in the light occupations incident to the care of the village and in sewing. The women seemed to have no special occupation outside of the very small amount involved in the care of the house.

The food consisted mainly of rice and lean beef, the former furnishing nearly seven-tenths and the two together nearly five-sixths of the total actual nutrients. In addition, chicken, fish, eggs, bread, green vegetables, and fruits were eaten in small quantities. So far as could be ascertained this did not differ very greatly from the home diet.

The investigations covered a period of 10 days during September, 1893. The food was carefully weighed as prepared for the table, and its composition in part determined by analysis and in part estimated.

DIETARIES IN THE UNITED STATES AND CANADA.

Massachusetts and Canada.—Nos. 84 to 115 were described on pages 145–156, above.

Connecticut.—In 1886 and later several dietaries of students, laborers, and one of a well-to-do private family in Middletown, were examined in connection with the study of Nos. 84 to 115. Nos. 117, 121, 122, 124, and 126 are selected from the group. The method employed is indicated in the text beyond. In general it was the same as that used in the Massachusetts dietaries, except that account was taken of the quantities of food rejected in the kitchen and table waste, and the amounts actually eaten were computed by subtracting these from the food purchased. In 1890 and later these investigations were renewed at Middletown in cooperation with the United States Department of Labor and as a part of the work of the Storrs Station. Nos. 116, 118 to 120, 123, 125, and 127 to 134 belong to the group thus studied. The method employed has been essentially that described on pages 200–204.

Dietaries of well-to-do private families in Middletown and Storrs.

No. 116 is the dietary of the family of a chemist in Middletown. During the month of April, 1891, the food actually eaten in the family of a chemist in Middletown was carefully weighed, and while no analysis was made of the food or of the table waste, it is believed that the figures given below are quite reliable. The family consisted of 1 man, a chemist, 34 years old and weighing 230 pounds, his wife, and a maidservant, who were at home during the entire month; also 1 man, a chemist, 27 years old and weighing 175 pounds, and an elderly woman were at the table about one-half of this period. The food materials used were mostly of such a nature that their composition is fairly well known from the averages of other analyses. The waste was reduced to a minimum, and it is probably fair to assume that no very considerable error was introduced by neglecting it. As both of the station chemists lived in the family and the wife of one of them was very much interested in the experiment, a very careful account of the weights of food used was kept. For the most part the food was weighed just before using, on a scale sensitive to 1 gram, with a charge of 10 kilograms.

No. 117, dietary of a professor's family in Middletown. The weighings and estimates for this dietary were made by Mr. Rockwood, assistant in the chemical laboratory of Wesleyan University. On the 1st day of January accurate account was taken of all the food materials in the house. The quantities of food brought to the house during January, February, and March were estimated from the grocer's, butcher's, and other bills. During the same period all the food left unconsumed, i. e., the kitchen and table refuse, was carefully kept, weighed, and its composition estimated. On the 1st day of April the food materials remaining in the house were weighed. In estimating the quantities of nutrients the method employed for those of the Massachusetts and Canadian dietaries was used.

The smallness of the waste is explained by the fact that the mistress of the house was a particularly careful housekeeper.

Nos. 118 and 119 are the dietaries of a professor at the Storrs Agricultural College. In the dietary made in the winter of 1893 the family consisted of the man, 32 years of age, his wife, and a maidservant; in that of the summer of 1893 the family was the same, except that the servant was absent and a child 1 year old included.

No. 120 is the dietary of the family of a retired jeweler. The family consisted of the father, 70 years of age, at very light work; his wife; their son, a chemist, 23 years old, with moderate exercise, and their daughter. The study was made during the month of September, 1891.

Dietaries of college and professional students in Middletown.

Nos. 121 to 123, three dietaries of a college students' club. A large number of the students in Wesleyan University board in clubs. The club, which may have any number of members up to 30, chooses one of its number as steward, and arranges with a matron to cook and serve the food which he purchases. Many of the members have to pay their way through college; the majority are obliged and the rest are content to have the cost of their board made low, even at the sacrifice of delicacies. While their diet is substantial and wholesome, they regard it as plain and economical. They are mostly from the Eastern States, and coming from the class of families whose sons go to college, it seems fair to assume that their habits of eating formed at home would not differ materially from those of the more intelligent classes of people in that part of the country. While the habits of many are sedentary rather than active, they nevertheless take considerable muscular exercise. In this respect they are very much like the students of other Eastern colleges. They are given to athletic sports in pleasant weather. Out of 250, sometimes 70 or more may be seen at once on the campus playing tennis and baseball. They could hardly be credited with as much muscular exercise on the average as laboring men doing moderate work, and they would therefore, without doubt, require somewhat less of protein as well as of the other nutrients in their food. Their requirements doubtless approach closely to those of the man with moderate exercise.

From the accounts of one of these clubs for a period of 3 months in 1886, the amounts of the several kinds of food materials purchased and the quantities of nutrients were computed. The results are given in No. 121 of the table.

The figures thus obtained represent what the students paid for, rather than the amounts actually consumed. The steward and some of the members of the club were of the opinion, however, that the amount of waste—that is to say, the material thrown away—was very small. It costs too much. But on investigating the matter more closely it appeared that a portion of the material served was left upon the plates and found its way into the garbage barrel or was given away. The rejected food was, therefore, collected during 1 week and weighed. Its composition was estimated and the amount of waste calculated.

The following term an examination of the dietary of the same club was made by Mr. Videon. Another steward was then in charge. He had learned of the excessive amounts of food in the former dietary, and planned to reduce the quantities. This was done largely by diminishing the meats. He states that he did not apprise the club of the change, and that it was not noticed. As he put it, "The boys had all they wanted, and were just as well pleased as if they had had more." The quantities of nutrients in both dietaries were estimated by the method already described. The waste in the second dietary was estimated with some care. In this, as in the first, it is assumed that the difference between the food purchased and the waste represents the actual consumption.

In the spring of 1893 the dietary was examined for a third time (No. 123). The observations were much more accurate, that is to say actual weighings and analyses were made of the food consumed.

No. 124 is the dietary of a second students' club. The members of the club were in moderate circumstances and felt the necessity of economizing, but at the same time intended to have an abundance of wholesome food. The observations were made by one of the members who was an advanced student in the chemical laboratory and much interested in these studies. The food materials were actually weighed and also the waste. The attention of the club had been called to the waste in the dietaries of the other clubs. In all probability this is the explanation of the smallness of the waste here.

No. 125 is the dietary of the members of a boarding club in a divinity school. The details are not yet published. The method was essentially that described beyond on pages 201–213.

No. 126. dietary of a college football team. This dietary was examined at a time when the team was in active training and boarding in a club by themselves. The observations were made at the same time and by the same person (Mr. C. S. Videou) as those of Nos. 122 and 124. Their exercise was vigorous and at times severe, but the examination was made near the close of the football season, when, in the judgment of members of the club, they were eating rather less heartily than they had done earlier in the season.

Mechanics, etc., in Middletown.—No. 127 is the dietary of a boarding house. The dietary commenced with supper, October 20, 1890, and continued until after dinner of November 19, a period of 30 days. During most of the time the family consisted of 13 men and 7 women. It very rarely happened that all of the family took all three meals at the house any given day. There were occasional visitors, and in this way once or twice the total number of meals taken per day was larger than the family alone would have required. The sex, approximate age and occupation of each member of the family, as it was constituted most of the time, were as follows:

Men:	
Machinists, 30 to 40 years of age	4
Machinist, about 35, after October 25	1
Harness maker, about 70	1
Hired men about the stable, one old, the other middle aged	2
Proprietor of the house, about 70	1
Manufacturers, one about 60, the other about 30	2
Chemist, about 27	1
Reporter for newspaper, about 20, after October 27	1
Total	13
Women:	
Housekeeper, about 30	1
Cook, about 45	1
Table girl, about 20	1
Doing no manual labor, about 30, 55, 55, and 70	4
Young lady at house 4 days, doing no labor	1
Total	8

Of the 13 men, 3 were counted as "hearty eaters," and 6 more as having decidedly good appetites.

No. 128 is the dietary of the family of a machinist, consisting of the father, born in Germany, 50 years of age, his wife, and 3 daughters, from 14 to 20 years of age. The period of observation was 30 days during November and December, 1891.

No. 129 is that of the family of a blacksmith and consisted of the father, a Canadian, about 40 years old, his wife, and 2 boys, 8 and 10 years of age. The study was made at the same time as No. 128.

Nos. 130 and 131 are 2 dietaries of the family of a stone mason. The father, born in Sweden, was about 28 years of age; the mother was also a Swede and had been in the United States only about 3 years; the other member of the family was a child 7 months old. The first study was made in November and December, 1892, and continued 28 days; the father was at work during 3 of the 4 weeks. The second study was made in April, 1893, under approximately the same conditions as the former one.

No. 132 is the dietary of the family of a carpenter. The family was of American birth and consisted of the father, who was about 25 years of age, the mother, and a boy 2 years of age. The study was made in November and December, 1892, and covered a period of 28 days.

Nos. 133 and 134 are 2 dietaries of another carpenter's family, also American, and consisting of the father, about 35 years, his wife, and a boy 11 years of age. The first dietary covered a period of 28 days in November and December, 1892, and the second a period of the same length in April and May, 1893.

Pennsylvania.—No. 135 gives the average minimum and maximum of 25 families in the poorest part of Philadelphia. With a single exception all of these families[1] were either foreign or of foreign descent.

Illinois.—No. 136 gives the average minimum and maximum of 26 families in the poorest part of Chicago.

Nos. 135 and 136 were made in 1892-93 by Miss Amelia Shapleigh, Dutton Fellow, of the College Settlements Association. Complete and exact accounts of food bought and eaten by 55 families in Philadelphia and Chicago, representing as many nationalities as possible and selected at random from the neighborhood of the settlement, were collected. The estimates ("amounts of food actually bought and put on the table") were made "after close questioning, observation, and in case of food not bought by the pound, careful weighing." Ten per cent of the total was deducted for rejected and undigested residues. No analyses of food materials were made and the basis upon which the amounts of nutrients was estimated is not stated.

United States Army and Navy rations.—It is difficult to compute exactly the food consumption by soldiers in the United States Army, because the men have more or less of opportunity to select or add to their food by commuting purchase or otherwise, so that the food actually used varies more or less from the regulation rations. The best data for calculations I have found are those contained in Circular No. 8 of the Surgeon-General's Office, United States Army. From these Mr. Woods has computed the nutrients and energy as stated in Nos. 137 and 138.

SUGGESTIONS REGARDING STUDIES OF DIETARIES.

The following statements are gathered from the experience of the writer and his associates, supplemented by their reading of the reports of the work of others.

The object of a dietary study is to determine the kinds and amounts of food materials and of nutriments consumed by one or more persons during a given period and under known and definitely stated conditions. The first essential is accuracy in the collecting of the original data. If these are indefinite, unreliable, or incomplete, the defects can not be compensated by the most accurate and painstaking chemical analysis, by elaborate treatment of the results, or even by a large number of observations. This accuracy is needed even in the smallest details; no part of the study of a dietary calls for more thought and care than the collection of the statistics regarding the food materials consumed, the people who consume them, and the proportions actually

[1] 1 American, 3 Irish, 6 German, 6 Jewish, 3 Italian, and 6 Negro.

eaten and not eaten. When these preliminary data have been properly collected, and the specimens have been secured for analysis, the work up to the writing of the results is largely a matter of routine which can be carried out in any well-ordered laboratory which is equipped for this kind of investigation. The reporting and especially the interpreting of the results requires not only an understanding of chemistry and physiology, but also an appreciation of some of the fundamental facts and principles of economics and sociology.

The philosophical study of food consumption requires the skill and the training of the chemist, the physiologist, the statistician, and the sociologist.

I. *Qualifications of the observer.*—For these reasons the person who collects the statistics should have a thorough scientific training and a clear idea of the questions involved and the importance of the task. Besides the technical knowledge, a goodly amount of tact is indispensable and the best success requires a certain sympathy with the people among whom the work is done, especially when they are in moderate circumstances or belong to the poorer classes. Food and household expenses are delicate subjects. A housewife or mistress of a boarding house may object to having her kitchen and account book subjected to critical examination unless it is done with tact. Perfect frankness and an explicit statement of the nature of the statistics, at the beginning of the study, is the safest rule. Often the offer of a small sum of money in compensation for the trouble caused by the inquiry will materially help the observer. The questions, often very direct, which the observer is obliged to ask, and the close watch which he must keep over even the smallest details of food purchased and wasted during the course of the investigation, are liable to place him in situations where deference is needed, for the good will and interest of the family must be kept at any cost, inasmuch as the observer is entirely dependent upon the subjects of his investigation for so much of his statistical information. Whether the observer had better be a man or a woman depends upon the individual and the circumstances of the case.

II. *Selection of persons for study.*—Whether the dietary to be studied is that of an individual (man, woman, or child), of a family, or of any other group of persons, as a boarding house or club of people of like class and occupation, the person or persons should be typical of a class, and normal in respect to health, development, and age.

(*a*) *Health.*—Families or groups containing invalids are not appropriate for the study of normal dietaries; neither should persons whose occupations or general habits exert an abnormal influence upon their dietary habits be selected, except in cases where special studies of the effect of such conditions are to be made.

(*b*) *Development.*—Greatly oversized or undersized persons should not be selected or included, and when children (either in families or individual experiments) are included or selected, height and weight as well as age should be carefully noted.

(c) *Age.*—Families or groups containing extremely old persons are hardly desirable. Studies of the food consumption of children or of aged people are best made with such persons by themselves.

Other conditions being the same, the greater the number of persons included in a dietary study the more valuable are the results as representing a class. There is, however, a limit in this direction which is set by the amount of labor which such a study involves. Accuracy and thoroughness must never be sacrificed, and the dietary of 1 person thoroughly studied is worth much more than that of a group of 20 or 30 studied indifferently. The number of imperfect studies is already considerable, and although they are valuable for the preliminary reconnoissance, the demand now is for a more accurate form of inquiry, which shall bring to light exact and not merely approximate facts. An experienced investigator with good laboratory facilities and plenty of assistance for the analyses, and especially for the preparation of samples of food materials and wasted food, may undertake the study of the dietary of a club of 25 or 30 persons, but it would be hardly advisable to undertake so large a study, even with all needed help, until considerable experience had been gained in the study of dietaries of individuals and families. The errors of my own early experience are the reason for the emphasis of this statement.

III. *Length of period of observation.*—One of the most serious criticisms to be made of a number of otherwise admirable dietary studies is that the period of observation has been too short. The demands of the organism for food may vary considerably from day to day. Changes in occupation, differences in the temperature and in the state of the weather in general, and minor ailments resulting from fatigue, exposure, or mental condition have a considerable effect upon the appetite; so that differences from one day to another may amount to a considerable portion of the total quantity of food consumed. For example, in a recent Italian dietary (No. 63 of Table 37) the period of observation covers 3 consecutive days, during 2 of which the subject was at work, while the third was a day of rest. On one of the working days he ate one-fourth more than on the other, and one-third more than upon the day of rest. This is by no means an extreme or exceptional case. For this reason studies of dietaries, either of individuals or families, covering a period of less than one or two weeks are not to be recommended unless they are to be frequently repeated, as was done by Camerer (see page 192). One month is a much more desirable length. Even for larger groups the time of observation should not be much less, inasmuch as, other conditions being the same, the longer the period of observation the more reliable the result.

IV. *Food—1. Collection and measurement of food materials.*—At the beginning of the study of a dietary a careful inventory should be taken of food materials[1] in stock in the kitchen, pantry, and cellar. The materials should in all cases be weighed rather than measured, and this applies not only to such materials as meats and flour, but also to such as milk, molasses, potatoes, beans, and other vegetables. Like accurate weighing should be made of all food materials purchased during the period of observation. Each should be carefully weighed when received; it is not safe to trust to purchase weights; in the case of meats, for instance, the cut may be trimmed at the market after weighing. At the end a second inventory should be made with the same precautions as observed at the beginning. The sum total of the food materials at the first inventory and of those received during the study, less the amount indicated by the final inventory, shows how much has been used during the period of observation.

2. *Description of food materials.*—The description of the various food materials, including that of each portion, e. g., each cut of each kind of meat, should be detailed, whether a sample is taken for analysis or not. The economic value would be increased by statements of price and quality. This is particularly to be observed in case of the meats; not only the kind, but the exact cut should be mentioned.

3. *Sampling for analyses.*—Samples of all food materials which form any considerable part of the dietary should be analyzed. This list usually includes all the meats, samples of which should be taken with each individual purchase, the lard, milk, butter, flour, oatmeal, and other vegetable foods, including canned vegetables and sugar.

The procuring of representative samples of meats is less simple than one without experience would suppose. To obtain fairly representative samples of steaks, chops, roasts, or ham is comparatively easy, as alternate or adjacent slices are readily procured at the time of purchase, but in the case of the miscellaneous cuts which form such a large proportion of the meat used in many kitchens, the problem of getting fair samples is often very perplexing. The matter is, however, one of foremost importance, and the only safe rule is to obtain samples at each and every purchase. To illustrate by a concrete example: In New England the custom of serving veal and mutton stews is very common. The meat selected for this purpose consists usually of the odds and ends of veal or mutton which happen to be in the market at the time, an assortment which it is impossible to duplicate. The best expedient we have found is to have these miscellaneous pieces cut into smaller ones—as small as possible without injuring the appearance of the meat—and

---

[1] In the studies of dietaries certain articles of food and drink are not, as a rule, included in the category of food materials. They are such as tea, coffee, salt, spices and condiments in general, including beef extract. Beverages containing nutritive material, e. g., beer, should be taken into account with food. Alcohol when consumed in small quantities, should be included (1 gram alcohol is approximately equivalent to 1.7 grams carbohydrates in fuel value); if alcohol is taken in large quantities the subject would not be fitted for observations upon normal nutrition.

to select for analysis pieces representative of the whole lot as regards proportion of meat and bone and lean and fat.

The practice of buying meat from the butcher's cart, instead of at the market, is a common one in many localities and makes the getting of samples for analysis very difficult, unless previous arrangements are made by which the piece of meat purchased shall be large enough for the use of the family and for the sample to be analyzed. The observer should always be at the market or meat wagon when purchases are made if this is possible, and must in any event attend to the taking of the sample for analysis whether at the market or in the kitchen; the market is generally the most convenient place for taking the sample.

Samples of flesh of large fish, such as halibut, salmon, etc., may be taken in the same manner as samples of meats, but with smaller fish which are sold by the piece and likewise with shellfish and also poultry and eggs, one or more samples similar in size and general appearance to those purchased for food should be selected for analysis.

Milk should be thoroughly mixed before sampling, and butter and cheese should be sampled as recommended by the Association of Official Agricultural Chemists.[1]

Lard may be sampled in the same way as butter.

The obtaining of representative samples of vegetable foods, such as flour, cereals, and vegetables, is comparatively easy, as is also the case with canned foods, whether animal or vegetable. Too much care, however, can not be taken to make sure that the samples for analysis shall be representative, as otherwise the accuracy of results will be greatly diminished.

4. Analyses.—Meats vary so much in proportion of refuse and edible portion and in the composition of the edible portion that it is necessary to analyze every piece if the work is to be exact. Indeed, we have felt compelled in late dietary studies to chop all the meat and take a small sample of each lot for analysis. This, of course, interferes with the cooking, but it makes more accurate sampling possible. With fish and poultry the necessity of repeated analysis is perhaps not so great. Milk should be analyzed very frequently, but the number of analyses may be considerably lessened by combining several samples taken upon different days. Butter should be analyzed at each purchase. Granulated sugar is usually free from moisture and mineral matter, but the moisture content of soft sugars and of molasses should be determined.

5. Waste.—From the economic standpoint the waste of food in American households is a serious matter, and it is desirable that exact statistics should be obtained and published. In the dietaries which we have studied the quantities of actual nutrients which have found their way to the garbage barrel have frequently amounted to a tenth, and in one

---

[1] See proceedings of the annual conventions of this association. U. S. Dept. Agr., Div. Chem.

case rose to a sixth of the whole food purchased. The real waste was worse than these figures imply, because the rejected material came very largely from animal food in which the nutrients are more costly than in the vegetable food materials. In the reports of dietary studies by the writer and his associates the term "waste" has been applied to all of the so-called "edible portion" of food which may for any reason be rejected. It includes all uneaten residues of the cooked food, e. g., meats, vegetables, bread, cake, and pastry, except, of course, the bone and other parts of meats and fish which would be classed as "refuse" in food analysis. It also includes all portions of food materials which are rejected before cooking, except those which belong to the refuse. Such materials as the parings of potatoes and turnips are reckoned as belonging to refuse, and are not included in the waste; they are, however, weighed, and their weights compared with the whole weight of the food material in order to determine the proportion of refuse.

Collection of waste.—The collecting of the waste is always difficult, as there are so many avenues through which it may be lost, and unless the observer can personally superintend its collection at each meal, the most explicit directions for the care of the garbage barrel and soap-grease bucket are necessary. When once collected the waste food materials should be carefully looked over, and bones (from which all adhering meat must be removed) and all hard substances, such as fruit pits, etc., which may have been accidentally included, should be removed. The waste is then to be dried for analysis.

Sampling.—The waste when thus partially dried is exposed to the air of the room for 28 to 48 hours and then weighed, chopped into pieces about the size of a pea, and carefully sampled. This sample, which should weigh about a kilogram, is very finely chopped, and a final sample of the customary size taken for the actual analysis. When the amount of waste is very large samples should be taken comparatively often, but in the case of an ordinary family the waste food for a period of 4 weeks will not present any special difficulty in obtaining a representative sample. When several samples are taken they may, of course, be combined, if care be taken that the amount of each particular sample which enters into the composite one shall be proportional to the quantity of waste which it represents.

Analysis.—The waste usually contains little crude fiber and only determinations of moisture, nitrogen, ether extract, and mineral matters are necessary.

6. *Digestibility.*—The ideal dietary study would include a digestion experiment. That is to say, the solid excreta from the food would be collected, weighed, and analyzed. This has been done in some of the recent work in Italy, Germany, and Japan.[1] Work in this direction in the United States is much to be desired.

---

[1] See detailed account of Nos. 44, 61 to 65, and 78, and special description of studies by Manfredi, pp. 173, 188, 191, 194.

V. *Data as to persons, length of period of observation, and climate.*—The most important of these data are the sex, age, weight, and occupation of the persons and the length of time during which their food consumption is observed.

1. Sex.—The number of persons of each sex must be noted. The estimation of comparative food consumption of men, women, and children and the number of men to which the number of women and children would correspond is explained beyond.

2. Age.—The exact age of each person, and especially of each child, should be recorded.

3. Weight.—The weight of each individual, or the average of all, should be noted—this especially to be borne in mind with children, statements of whose development, both weight and height, are also desirable.

4. Occupation.—The occupation of each person should be stated in detail. The relation between food consumption and the kind and amount of work done is as yet too little understood, and data bearing upon it much to be desired. It would obviously be misleading to compare the dietaries of professional men with those of men at muscular labor, and those of men at muscular work of different degrees of severity, without taking the nature of the labor into account. Of course these estimates of the severity of different kinds of labor can not be definite, but they should be described as accurately as possible.

5. Time of observation.—The exact times during which the food consumption of each person is observed must be carefully noted. This is most easily done by keeping an exact account of the number of meals eaten by each person during the period of observation. The number may be expressed in terms of days by dividing the total number of meals by the number of meals per day. This procedure is to be recommended, because members of the family or other groups are apt to be absent more or less from meals, and other persons are often present at meals.

6. Climate, season, temperature.—If the people spend their time in well-warmed houses, differences of climate and season may perhaps have little influence upon food consumption. Otherwise these factors may be most important. But little is certainly known as to their actual effect, and data are most desirable.

7. Supplementary statistical and sociological data.—In addition to the statistical data above referred to, certain other information is needed to give the results their full meaning. Nationality, home life, environment, health, income, and expenditures are among the data to be especially considered.

VI. *Calculation of results.*—The statistics of food consumption in dietaries are usually reduced to terms of nutrients consumed per man per day. Data as to total amounts of food materials of different kinds and of food wasted and the composition of each are obtained in the manner mentioned above. From these we may calculate the total amounts of protein, fats, and carbohydrates. Dividing these quantities by the total number of days' food for 1 person gives the quantities *per person* per day. This, however, is not sufficiently accurate, for it is necessary to distinguish between persons of different classes. It would be obviously unfair to compare the amounts of nutrients consumed by a child 5 years of age with those consumed by an adult man. It is therefore desirable to express the food consumption of men, women, and children in equivalents of per man per day. To do this we must find the ratio which the food consumption of children of various ages and of women bears to that of the man.

The ratios used in the dietary studies by the writer and his associates are the same as were explained on page 148. It may prove desirable to change them later. I hope to revert to this in a future discussion of standards for dietaries. Dietary habits are numerous and complex, and a dietary studied without taking these elements into account is thereby deprived of its real significance.

CLASSES OF PEOPLE AND LOCALITIES FOR DIETARY STUDIES.

The studies which have so far been made are limited in locality to several towns and cities in Massachusetts and Connecticut, and to Philadelphia and Chicago. In New England these studies include dietaries of professional men and students, factory operatives, mechanics, such as blacksmiths, carpenters, masons, and machinists, and also various classes of less skilled laborers, such as brickmakers, etc. They are very likely, so far as they go, typical of the town population of the New England States. In Philadelphia and Chicago only the food consumption of families of the poorer classes of society, mostly foreigners, has been studied.

It now seems desirable that these studies should be extended in two ways. They should include a number of classes of people and cover a larger territory. They can be advantageously carried on with the aid of colleges and experiment stations, although other organizations, as labor bureaus and benevolent societies, might most advantageously cooperate.

Among the classes that need to be studied are: First, people in professional and business life; those whose labor is intellectual rather than muscular. These would include the families of business and professional men, and likewise students. Second, farmers and their families in the country districts. Third, mechanics of different classes, and with muscular work of different degrees of severity, as carpenters, masons, blacksmiths, etc. Fourth, factory operatives. Fifth, ordinary laborers of various classes. Sixth, people of the poorer classes in the large cities. Seventh, inmates of hospitals and prisons. Eighth, children of different ages.

CHAPTER X.

STANDARDS FOR DIETARIES.

Various standards have been proposed by physiologists and chemists for daily dietaries for persons of different age, sex, and occupation, and in different conditions of life. They are usually estimated in terms of the three classes of nutritive ingredients, protein (or albuminoids), fats, and carbohydrates.

Our best information regarding dietary standards has come from Germany where studies have been made by numerous investigators, such as Liebig and especially Voit and his followers of the Munich school of physiologists. Voit's standards are the ones most commonly followed in estimates of dietaries. Outside of Germany the names of Playfair in England, Payen in France, and Moleschott in Italy, deserve especial mention as contributors to the knowledge of the subject. It is a noteworthy fact, however, that very little attention appears to have been paid in either the United States or England to the results of the latest and best research in this direction. Even the text-books on chemistry and physiology in the English language which are looked upon as most authoritative are apt, when they treat of the matter, to do so most superficially.[1]

Unfortunately, in the treatment of this subject there is much confusion as to the real significance of the term dietary standard and the basis upon which the standards may be estimated. The same confusion obtains in still greater degree in the discussions of standards of rations for domestic animals.

METHODS OF ESTIMATING DIETARY STANDARDS.

A standard for a ration or a dietary may be based upon either:

(1) The observed facts of consumption. Thus the standards of Playfair, Moleschott, and Voit for a laboring man at moderate work are based mainly upon the quantities actually consumed by persons whose food consumption they had learned from their own or other observations.

---

[1]The subject is well handled in a number of German works including the following, in which references to the original investigations may be found:

Voit. Physiologie des allgemeinen Stoffwechsels und der Ernährung, Vol. V, of Hermann's Handbuch der Physiologie, 1881.

Foster. Ernährung und Nahrungsmittel in Vol. I, of Pettenkofer and Ziemssen's Handbuch der Hygiene, 1882.

König. Chemie der menschlichen Nahrungs- und Genussmittel, 3d edition, Vol. I, 1891.

Hammersten. Physiologische Chemie, 1891.

(2) The actual need and available or most economical supply. Standards estimated on this basis would take into account not only the actual demands of the body for nourishment, but also the kinds of food that are to be had and the pecuniary cost of the nutrients in different food materials.

(3) The physiological demand. By this is meant the quantities of nutrients which are most appropriate for the particular individual or class. The basis here would be found in the actual facts of normal metabolism. The data of this sort now at hand are too few to make estimates reliable.

Standards estimated upon the first basis would correspond to actual practice. Those upon the second would take into account both pecuniary and hygienic economy. Those of the third would consider only what was intrinsically most fitting.

The current standards for daily dietaries for men and daily rations for domestic animals have been based mainly upon the considerations in the first class. In the feeding of domestic animals and in the nutrition of people who are limited in respect to either the supply of food to which they have access or their means for purchasing their food the data of the second class would be appropriate. But for people in moderate and comfortable circumstances in most parts of Europe and the United States, where commerce brings a large variety of food materials to every market and incomes are large enough to warrant use of properly selected food, data of the third class are the ones to be taken into consideration.

EUROPEAN AND AMERICAN DIETARIES AND STANDARDS.

The pages which precede have outlined, though very incompletely, the data now available regarding the actual food consumption, and have indicated in a general way those which have been thus far obtained by experimental inquiry regarding the physiological demand. I hope, in a future publication, to discuss this subject more fully and to indicate what appear to be the inferences to be derived from the knowledge now at our disposal and what are the lines of inquiry which are needed to make that knowledge more nearly adequate.

Meanwhile, I venture to quote the following from a previous article upon the subject.[1]

It will be observed that this discussion is based upon the assumption that the body requires for its nourishment—

(1) Enough of protein to make up for the protein and muscle and other nitrogenous substances consumed in the body.

(2) Enough energy to supply the demand for heat and muscular work. The estimates for energy are based upon Rubner's factors (4.1 calories for each gram of protein, the same for each gram of carbohydrates,

[1] Report of Storrs (Conn.) Experiment Station for 1891, pp. 145-149.

CHART 3.—DIETARIES AND DIETARY STANDARDS.

Quantities of nutrients and energy in food for man per day.

[Amounts of nutrients in pounds. Fuel value in calories.]

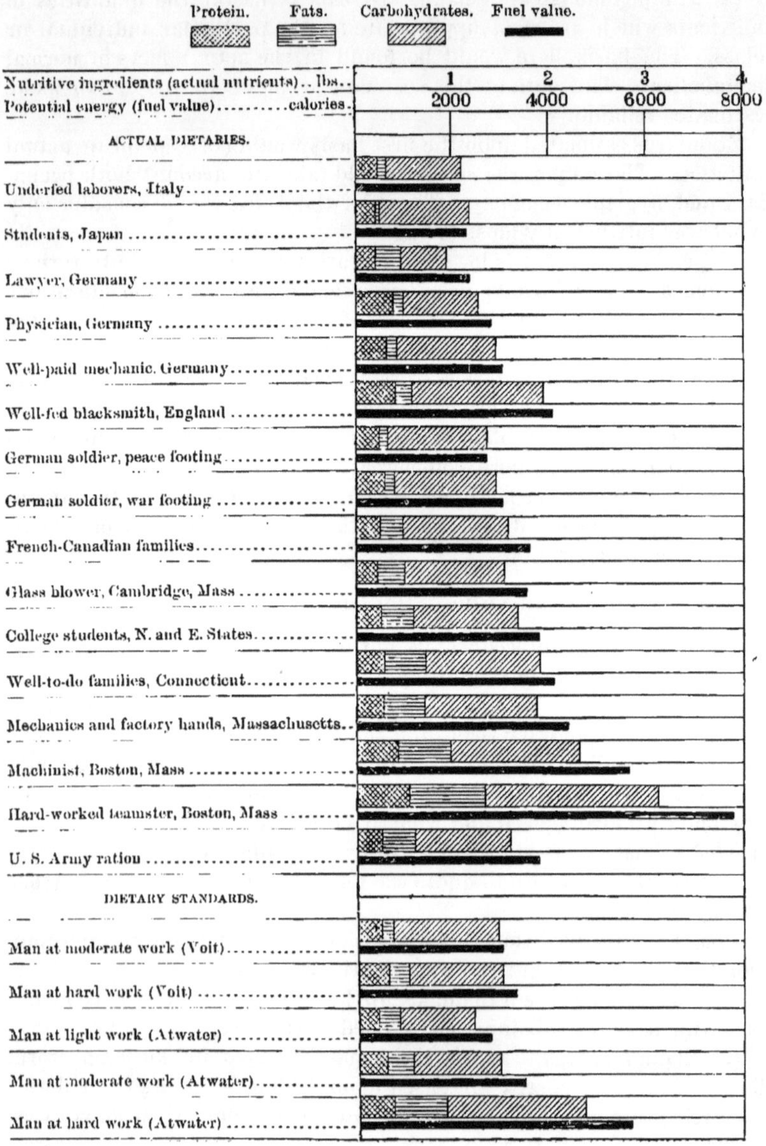

and 9.3 for each gram of fats). It is further assumed that in serving as fuel the fats and carbohydrates may replace one another in proportion to their potential energy, so that it is not necessary to lay down specific quantities of either, provided the sum of the two is sufficient for the purpose. The fitting of the food to the demands of the palate and the digestive apparatus are, of course, factors of the utmost importance, but can not be discussed here.

European dietary standards.—The standards in Table 38, compiled from European sources, are intended to represent roughly the needs of average individuals of the classes named in England and Germany. Nos. 1 to 8 are based upon observations of Voit and his followers. For Nos. 1 and 3 the results of numerous estimates by Dr. Camerer of the food of his own children, and by Voit and Forster of the food consumed by a number of other children, are used. No. 2 was calculated by myself from the data of Nos. 1 and 3. The figures given represent means between smaller quantities for the younger and larger for the older children. The data collated in Table 37 above seem to call for a revision of the standard for children as well as those for adults. Nos. 5 and 6 are based upon observations chiefly by Forster. Nos. 7 and 8 are by Voit, and based both upon quantities consumed by individuals under experiment, and upon observed dietaries of a much larger number of persons in Germany. No. 7 is the one so frequently quoted as Voit's standard for a "mittlerer Arbeiter," a laboring man at moderate work.[1] These figures represent the average needs of ordinary mechanics and laboring men at their ordinary work, as estimated from the food consumed by such men in Germany and especially in the region of Munich. For men not engaged in muscular labor Voit regards it as better to use less carbohydrates, not over 350 grams per day, and supply the remaining need with fat. In other words, the food of well-to-do people, with intellectual rather than muscular labor, may advantageously contain more meat, butter, and other animal foods rich in fat, than the largely vegetable food of the German working people supplies. Voit would have somewhat over half of the 118 grams of protein of the food of the average laboring man at moderate work supplied by meat and other animal foods.

It will be borne in mind that these quantities, like those in the tables of this article, generally refer to the total rather than the digestible nutrients of the food. Such dietaries as Voit proposes would contain approximately the following amounts of digestible nutrients:

	Protein.	Fats.	Carbohydrates.
	Grams.	*Grams.*	*Grams.*
For laboring man at moderate work	116	53	450
For laboring man at hard work	130	95	402

No. 9 is by Moleschott, whose earlier scientific life was spent in Germany, but who has lived for many years in Italy. It may therefore be called either Italian or German. No. 10 is German. Nos. 11 to 15 are dietaries estimated by Playfair, an Englishman, from the data cited in Table 37 and others.[2] The figures for No. 11, subsistence diet, are based upon dietaries of reconvalescents in hospitals, of people in prison, and of laboring people in England, and in the so-called "cotton famine" of 1862-63. For No. 12, diet in quietude, the dietaries of soldiers in time of peace, and for No. 13, diet of adult in full health, army dietaries (English and continental) in war were employed. The estimates for active laborers, No. 14, were based upon

[1] Physiologie des Stoffwechsels und der Ernährung, 519-523.

[2] Chem. News, XI, 1865, 222.

observations of the food of men at moderately active work, who would correspond in demand for nutrients to the "mittlerer Arbeiter" for which the dietaries of Voit, Moleschott, and Wolff, Nos. 7, 9, and 10, were calculated. The figures of No. 15 were based upon observations of food of men in England at harder work, such as those of Nos. 7, 8, and 9 of Table 38.

TABLE 38.—*European standards for daily dietaries for people of different classes.*

Reference No.		Nutrients.				Potential energy.
		Protein.	Fats.	Carbo-hydrates.	Total.	
	Children:	*Grams.*	*Grams.*	*Grams.*	*Grams.*	*Calories.*
1	1 to 2 years, average	28	37	75	140	765
2	2 to 6 years, average	55	40	200	295	1,420
3	6 to 15 years, average	75	43	325	443	2,040
4	Aged woman	80	50	260	390	1,860
5	Aged man	100	68	350	518	2,475
6	Woman at moderate work	92	44	400	536	2,425
7	Man at moderate work (Voit)	118	56	500	674	3,055
8	Man at hard work (Voit)	145	100	450	695	3,370
9	Man at moderate work (Moleschott)	130	40	550	720	3,160
10	Man at moderate work (Wolff)	125	35	540	700	3,030
11	Subsistence diet (Playfair)	57	14	341	412	1,760
12	Diet in quietude (Playfair)	71	28	341	440	1,950
13	Adult in full health (Playfair)	119	51	531	701	3,140
14	Active laborers (Playfair)	156	71	568	795	3,630
15	Hard-worked laborers (Playfair)	185	71	568	824	3.750

It is hardly necessary to explain that the standards of this table represent only general averages. Thus Voit, Playfair, and the other physiologists named assume that for an ordinary laboring man, doing an ordinary amount of work, the amounts of nutrients stated in Nos. 7, 9, 10, and 14 will suffice; that with them he will hold his own, and that any considerable excess above these quantities will be superfluous. No one expects any given man to adjust his diet exactly to either of these standards. He may need more, and may perhaps get on with less. He may eat more fats and less carbohydrates, or he may consume more protein, if he is willing to pay for it. If, however, he has much less protein and keeps up his muscular exertion, he will be apt, sooner or later, to suffer.

Of course, different individuals, under like conditions, will both require and consume different quantities of nutrients. In general, the larger the person, that is to say the bulkier the machinery in his organism, the more of protein and other nutrients will be consumed. Hence, men need, on the average, more than women and children. The requirements vary with the muscular activity. A man at hard work requires more nutrients than one at lighter work or at rest. Aged people, who are generally less active than those in the prime of life, require less food.

The four dietaries by Voit, Moleschott, Wolff, and Playfair, just mentioned (Nos. 7, 9, 11, and 14), have long been accepted by physiologists and chemists as expressing about the average quantities of nutrients which a man doing moderately hard, muscular work would need in his food each day. They vary considerably from each other, however. That of Moleschott, for instance, calls for 130 grams of protein; that of Voit, only 118. There are similar differences in the quantities of fat and carbohydrates. But no one adjusts his food exactly to chemical standards. Different people consume very different foods, and yet they get on very well, and it is perfectly clear that either of these standards may be right enough. And different as they are, a remarkable agreement between them has lately come to light.

When these standards were proposed experimental science had not taught how to measure the fuel value of food by the potential energy of its constituents. Late research has told how this may be done. The energy is measured in heat units, for which calories are here used. A gram of protein or of carbohydrates is assumed to contain 4.1 and a gram of fats 9.3 calories. Applying this measure to these dietaries, the extreme variation in the four is only from 3,032 to 3,160 calories. That is to say, four of the most prominent investigators, Playfair in England, and the others in Germany and Italy, working with different people and by more or less different methods, arrived at estimates which vary somewhat in the proportions of the nutrients, but when the different standards are reduced to terms of potential energy they agree almost exactly. The closer scientific scrutiny, which the latest and most painstaking research has made practicable, serves only to bring the apparent discrepancies into accord and thus confirm, in an unexpected and most striking way, the correctness of the standards.

To doubt the conclusions of such eminent authorities, when these conclusions are based upon such diversified experience and experiment and are substantiated in so striking a way as that just described, may seem presumptuous. I venture, nevertheless, to urge that these standards do not represent the quantities of nutritive material that the average mechanic or other workingman needs in order to do a fair day's work; that the allowance is too small for what such a man ought to do and can well do. The reasons for this view are found in the teachings of later experimental research, especially that of Voit, and others who have been associated with him, regarding the functions of food and its nutritive ingredients, and in the studies of American dietaries above described and the inferences which they seem to warrant. The kernel of the whole question is found in the fact that the European standards are based upon the food consumption of people whose plane of living is low in comparison with that of the people in the United States. The thesis which I attempt to defend is that to make the most out of a man, to bring him up to the desirable level of productive capacity, to enable him to live as a man ought to live, he must be better fed than he would be by these standards. This is only part of the story, but it is an essential part. The principle is one that reaches very deep into the philosophy of human living.

American v. European standards.—In the passage just quoted emphasis is laid upon the fact that the current European dietary standards are based mainly upon the facts of actual food consumption. Thus, Voit's standard for a laboring man at moderate work is based chiefly upon his observations of the quantities of food actually consumed by manual laborers, mostly mechanics, in Munich and other places in Bavaria, who were reasonably well paid and reasonably well fed, as judged by the standards of wages and living which obtained in these places at the times when the observations were made, twenty years or more ago. In the same way Playfair's estimates were based chiefly upon the conditions which he observed in England some thirty years ago. If either one had used such data as he would find in New England to-day he doubtless would have made his dietary standards correspond. The most cursory examination of Table 37 shows that the American standard in this respect would be much higher than the European. My own belief is that the American standard is a much more desirable one. The scale of living or "standard of life" here is much higher than it is in Europe. People in Massachusetts and Connecticut are better housed, better clothed, and better fed than in Bavaria or Prussia, they do more work, and they earn higher wages. While the statistics are not as full as is to be desired, there appears to be a consensus of those who have observed most closely that the facts lie in this direction. Very likely what Voit reckons as hard or severe muscular work would count here as only moderate work.

Considering the body as a machine, the American workingman has a more strongly-built machine and more fuel to run it than has his European brother. While it is not absolutely proven, it seems in the highest degree probable that the higher standard of living, the better nutrition, the larger product of labor and the higher wages go together. It is in view of such considerations as these that I have ventured to suggest more liberal standards for dietaries than those which have been proposed by the European authorities above quoted. The following standards are from the article above referred to:[1]

SUGGESTIONS FOR DIETARY STANDARDS.

In Table 39, I venture to suggest certain proportions of protein and energy which may be appropriate as averages for dietaries for people of different forms of activity. In the method by which the estimates are made, these agree with those of Voit and Playfair in that the proportions assumed to be needed are inferred empirically from those consumed in observed cases. They differ from the ones referred to in the use of the experience that has accumulated during the not far from two decades since the latest of the latter, those of Voit, were proposed; a period of no little activity of research in the science of nutrition. But the principal reason for the wide differences in the quantities is that the results of the inquiries regarding American dietaries have been taken into account in the estimates here given.

Concerning the quantities of protein, it must be confessed that the experimental data do not yet suffice for at all exact estimates even for average persons of different classes. The proportions here given are such as seem to me reasonable in the light of the present knowledge. The same may be said regarding the figures for energy.

The estimates in Table 39 are expressed simply in terms of protein and energy. There is really little ground for giving quantities of fats and carbohydrates in such standards, since the two are mutually replaceable, and the quantities must vary with the conditions of consumption.

Regarding the standards of Table 38, a few additional remarks will suffice.

It will be observed that the quantities are in a well-nigh regularly ascending scale. The lowest are such as have been found to suffice amply for men of sedentary life. The highest accord with the larger, but not the largest, of the American dietaries quoted above. The quantities in No. 1 are about the same as were found ample for the German professors and lawyers above referred to, who were men of decided mental activity. They might be small for the average man of like occupation with us, in which case such proportions as those of No. 2 would be more appropriate. The standard for a man at moderate work, which Voit places at 118 grams of protein and 3,050 calories, and

[1] Report of Storrs (Conn.) Experiment Station, 1891, p. 160.

Playfair and other authorities at very nearly the same figures, I have placed at 125 grams of protein and 3,500 of energy. These are smaller than the averages of the American dietaries, but I have assumed that the waste in the latter would count for considerable and that the quantities actually eaten would not be so very far above these figures, particularly as the allowance here for a man at active muscular work is 150 grams of protein and 4,000 calories of energy.

It has been assumed a woman requires on the average eight-tenths as much as a man for corresponding muscular activity. I have assigned the dietary of a man with light exercise to a woman at moderate work, and that of a man with very little physical exercise to a woman at light work, thus providing for a woman rather more than eight-tenths as much as a man. Very likely eight-tenths would accord more nearly with the actual needs.

It is quite possible that the quantities of nonnitrogenous nutrients necessary to furnish the energy called for in the following standards would be large in comparison with the amounts of protein; in other words, that the nutritive ratios are too wide for normal nutrition. In this respect they are a compromise between the currently accepted European standards and the actual dietaries observed in New England.

TABLE 39.—*Standards for daily dietaries (American).*

	Protein.	Fuel value.	Nutritive ratio.
	Grams.	*Calories.*	*1:*
Woman with light muscular exercise	90	2,400	5.5
Woman with moderate muscular work } Man without muscular work }	100	2,700	5.6
Man with light muscular work	112	3,000	5.5
Man with moderate muscular work	125	3,500	5.8
Man with hard muscular work	150	4,500	6.3

CHAPTER XI.

ERRORS IN OUR FOOD ECONOMY.[1]

Fortunately, enough information has already been gained to indicate in a general way what are the principal mistakes in the food economy of people in the United States, even though we are not yet certain as to all the details.

Scientific research, interpreting the observations of practical life, implies that several errors are common in the use of food:

First, many people purchase needlessly expensive kinds of food, doing this under the false impression that there is some peculiar virtue in the costlier materials, and that economy in our diet is somehow detrimental to our dignity or our welfare. And, unfortunately, those who are most extravagant in this respect are often the ones who can least afford it.

Secondly, the food which we eat does not always contain the proper proportions of the different kinds of nutritive ingredients. We consume relatively too much of the fuel ingredients of food, such as the fats of meat and butter, the starch which makes up the larger part of the nutritive material of flour and potatoes, and sugar and sweetmeats. Conversely, we have relatively too little of the protein of flesh-forming substances, like the lean of meat and fish and the gluten of wheat, which make muscle and sinew and which are the basis of blood, bone, and brain.

Thirdly, many people, not only the well-to-do, but those in moderate circumstances, use needless quantities of food. Part of the excess, however, is simply thrown away with the wastes of the table and the kitchen; so that the injury to health, great as it may be, is doubtless much less than if all were eaten. Probably the worst sufferers from this evil are well-to-do people of sedentary occupations—brain workers as distinguished from hand workers.

Finally, we are guilty of serious errors in our cooking. We waste a great deal of fuel in the preparation of our food, and even then a great deal of the food is very badly cooked. A reform in these methods of cooking is one of the economic demands of our time.

[1] The statements in this chapter are taken mainly from articles by the author in the Century Magazine, The Forum, and the Report of the Storrs (Conn.) Experiment Station for 1892.

CHEAP V. DEAR FOOD.

We can not judge of the nutritive value of food by the quantity. This fact is brought out clearly by the figures in Table 16 on page 139. There is as much nutriment in a pound of wheat flour as in 3½ quarts of oysters, which weigh 7 pounds. There is still less connecnection between nutritive value and price. In buying at ordinary market rates we get as much material to build up our bodies, repair their wastes, and give strength for work, in 5 cents' worth of flour or beans or codfish, as 50 cents or $1 will pay for in tenderloin, salmon, or lobsters.

Round steak at 15 cents a pound is just as digestible and is fully as nutritious as tenderloin at 50. Mackerel has as high nutritive value as salmon, and costs from an eighth to half as much. Oysters are a delicacy. If one can afford them there is no reason for not having them, but 25 cents invested in a pint would bring only about an ounce of protein and 230 calories of energy. The same 25 cents spent for flour at $6 a barrel, or 3 cents a pound, would pay for nine-tenths of a pound of protein and 13,700 calories of energy. When a day-laborer buys bread at 7½ cents a pound the actually nutritive material costs him three times as much as it does his employer, who buys it in flour at $6 a barrel.

FOOD OF THE POOR.

Illustrations of the prejudice of people, especially those in moderate circumstances, against the less expensive kinds of food are very common.

Mr. Lee Meriwether, who has given much attention to this special subject, cites a case in point, that of a coal laborer who boasted: "No one can say that I do not give my family the best of flour, the finest of sugar, the very best quality of meat." He paid $156 a year for the nicest cuts of meat, which his wife had to cook before 6 in the morning or after half past 6 at night, because she worked all day in a factory. When excellent butter was selling at 25 cents a pound he paid 29 cents for an extra quality. He spent only $108 a year for clothing for his family of 9, and only $72 a year for rent in a close tenement house, where they slept in rooms without windows or closets. He indulged in this extravagance in diet when much less expensive food materials, such as regularly come upon the tables of men of wealth, would have been just as nutritious, just as wholesome, and in every way just as good, save in the gratification to pride and palate. He was committing an immense economic blunder. Like thousands of others, he did so in the belief that it was wise and economical.

The sad side of the story is that the poor are the ones who practice the worst economy in the purchase as well as the use of food. The Massachusetts Bureau of Labor, in collecting the dietaries above referred to, made numerous inquiries of tradesmen regarding the food of the poor in Boston, meaning by poor "those who earn just enough to keep themselves and familes from want." The almost universal testimony was, "They usually want the best and pay for it, and the most fastidious are those who can least afford it." The costliest kind of meat, the finest flour, and very highest priced butter were demanded, and many scorned the less expensive meats and groceries such as well-to-do and sensible people were in the habit of buying.

I have taken occasion to verify these observations by personal inquiry in Boston markets. One intelligent meat man gave his experience with a poor seamstress, who insisted on buying tenderloin steak at 60 cents per pound. He tried to persuade her that other parts of the meat were just as nutritive, as they really are, but she would not believe him; and when he urged the wiser economy of using them she became angry at him for what she regarded as a reflection upon her dignity. "My wealthy customers," said he, "take our cheaper cuts, but I have got through trying to sell these economical meats to that woman and others of her class."

I am told that people in the poorer parts of New York City buy the highest priced groceries, and that the meat men say they can sell the coarser cuts of meat to the rich, but that people of moderate means refuse them. I hear the same thing in Washington and other cities.

ONE-SIDEDNESS OF OUR DIETARY.

I have said that our diet is one-sided, that the food which we actually eat has relatively too little protein and too much fat, starch, and sugar. In other words, it is relatively deficient in the materials which make muscle and bone and contains a relative excess of the fuel ingredients. This is due partly to our large consumption of sugar and partly to our use of such large quantities of fat meats. In the statistics of dietaries above referred to the quantities of fat in the European dietaries range from 1 to 5 ounces per day, while in the American the range is from 4 to 16 ounces. In the daily food of well-to-do professional men in Germany, who were amply nourished, the quantity of fat is from 3 to 4½ ounces per day, while in the dietaries of Americans in similar conditions of life it ranges from 5 to 7½ ounces in the food purchased. The quantities of carbohydrates in the European dietaries range from 9 to 24 ounces, while in the corresponding American dietaries the carbohydrates were from 24 to 60 ounces.

It is customary to estimate the proportion of fuel ingredients to protein in what is called the nutritive ratio. In this estimate 1 part, by weight, of fats is counted as equivalent to 2¼ of carbohydrates. Adding the two together gives the amount of the fuel ingredients. In the American dietaries the proportion of fuel ingredients to 1 part of protein ranges from 6.6 to 8.7, and even higher. In the European dietaries of well-nourished people and in the dietary standards which express the average needs according to the teachings of the best physiological observations it is from 4.1 to 6, or thereabout. The rejection of so much of the fat of meat at the market and on our plates at the table is not mere willful wastefulness, it is in obedience to nature's protest against a one-sided and excessive diet.

But the most remarkable thing about our food consumption is the quantity. The American dietaries examined in the inquiry mentioned above were of people living at the time in Massachusetts and Connecticut, though many came from other parts of the country. It would be wrong to take their eating habits as an exact measure of those of people throughout the United States. For that matter, a great deal of careful observation will be needed to show precisely what and how much is used by persons of different classes in different regions. Just this kind of study in different parts of the country is greatly needed. But such facts as I have been able to gather seem to imply that the figures obtained indicate in a general way the character of our food consumption. Of the over 50 dietaries of reasonably well-to-do people thus far examined the smallest is that of a mechanic's family. In this the potential energy per man per day was about 3,000 calories. The next smallest was that of the family of a chemist who had been studying the subject and had learned something of the excessive amounts of food which many people with light muscular labor consume. This dietary supplied 3,200 calories of energy per man a day. The largest was that of brickmakers at very severe work in Massachusetts. They lived in a boarding house managed by their employers, who had evidently found that men at hard muscular work out of doors needed ample nourishment to do the largest amount of work. The food supplied 8,850 calories per day.

Voit's standard for a laboring man at moderate work, which is based upon the observation of the food of wage workers who are counted in Germany as well paid and well fed, allows 118 grams of protein and 3,055 calories of energy. The standards proposed by myself in which the studies of American dietaries have been taken into account allow 125 grams of protein and 3,500 calories of energy for a man at moderately hard muscular work. The dietaries of Massachusetts and Connecticut factory operatives, day laborers, and mechanics at moderate work averaged about 125 grams of protein and 4,500 calories of energy. For a man at "severe" work, Voit's standard calls for 145 grams of protein and 3,370 calories of energy. The Massachusetts and Connecticut mechanics at "hard" and "severe" work had from 180 to 520 grams of protein and from 5,000 to 7,800 calories of potential energy, and in one case it rose to the 8,500 just quoted. In the dietary standards proposed by myself it did not seem to me permissible to assign less than 4,000 calories to that for a working man at "hard," and 5,700 for a man at "severe" work.

Just what compounds in food are required for the nutriment of the brain, physiological chemistry has not yet told us; but it is certain that people with little muscular exercise require less food than those at hard muscular labor. Many men whose work and strain are mental rather than physical suffer from overeating. In a number of dietaries of professional men in Germany, Denmark, and Sweden, including a university professor, a lawyer, physicians, and students, all of whom were in comfortable circumstances, in good health, and amply nourished, the energy varied from 2,180 to 3,035 calories; the mean was about 2,600 calories. The average of the dietaries of professional men and students from the Northern and Eastern States, residing in Middletown, Conn., was 4,155 calories. The range was from 3,205 in the family of the chemist to whom I have referred, to 5,345 in a students' boarding club. These figures, like the others of the American dietaries cited, refer to the food purchased. The average of the food eaten was 3,705 calories. In the students' dietary the food eaten supplied 4,825 calories.

Now it is not easy to see why these men required so much more than was sufficient to nourish abundantly men of like occupation, but unlike temptation to overeating, in Europe. Difference in climate can not account for it. We are a little more given to muscular exercise here, which is very well for us, but it cannot justify our eating so much. In the German army, where especial attention is given to diet, and it has been an axiom since the time of Frederick the Great, that soldiers to march well and fight well must be well fed, a ration for time of peace has been computed at 2,800, and one for time of war at 3,095 calories. During the last Franco-German war, shortly before the battle of Sedan, an order was issued by King, afterwards Emperor, William, which provided an extraordinary war ration, which is estimated at 3,985 calories. If a man with a tremendous physical and nervous tension, required in such terrible service as the German soldier was called upon to render in his victorious contests with the Frenchman, is well supplied by a ration of less than 4,000 calories of energy, and German professional men in their quiet but active and successful intellectual work at home are amply nourished with 2,700 calories and less, how happens it that men of mental rather than muscular occupation here consume food with 4,000 calories and more?

I think the answer to this question is found in the conditions in which we live. Food is plenty. Holding to a tradition which had its origin where food was less abundant, that the natural instinct is the measure of what we should eat, we follow the dictates of the palate. Living in the midst of abundance, our diet has not been regulated by the restraints which obtain with the great majority of the people of the Old World, where food is dear and incomes are small.

Indeed, the very progress which we are making in our civilization brings with it increased temptation to overeating. The four quarters of the earth are ransacked to supply us with the things which will most tempt our appetites, and the utmost effort of cooks and housewives is used in the same direction. It is all the more fitting, therefore, that information as to our excesses and the ways of avoiding them should come at the same time.

How much harm is done to health by our one-sided and excessive diet no one can say. Physicians tell us that it is very great. Of the vice of overeating, Sir Henry Thompson, a noted English physician and authority on this subject, says:

> I have come to the conclusion that more than half the disease which embitters the middle and latter part of life is due to avoidable errors in diet, * * * and that more mischief in the form of actual disease, of impaired vigor, and of shortened life accrues to civilized man * * * in England and throughout central Europe from erroneous habits of eating than from the habitual use of alcoholic drink, considerable as I know that evil to be.

This is in the fullest accord with the opinions of physicians and hygienists who have given the most attention to the subject, and these opinions are exactly parallel with the statistics here cited.

WASTE OF FOOD IN AMERICAN HOUSEHOLDS.

The direct waste of food occurs in two ways, in eating more than is needed and in throwing away valuable material in the form of kitchen and table refuse. That which is thrown away does no harm to health, and in so far as part of it may be fed to animals or otherwise utilized, it is not an absolute loss. That which we consume in excess of our need or nourishment is worse than wasted because of the injury it does to health. A few instances taken from the investigations mentioned above will help to illustrate the waste of food.

One of the dietaries examined by the Massachusetts Labor Bureau was that of a machinist in Boston who earned $3.25 per day. In food purchased the dietary furnished 182 grams of protein and 5,640 calories of energy per man per day, at a cost of 47 cents. One-half the meats, fish, lard, milk, butter, cheese, eggs, sugar, and molasses would have been represented by 57 grams of protein, 1,650 calories, and 19 cents. If these had been substracted the record would have stood at 125 grams, 3,990 calories, and 28 cents. This family might have dispensed with one-half of all their meats, fish, eggs, dairy products, and sugar, saved 40 per cent of the whole cost of their food, and still have had all the protein and much more energy than is called for by a standard which is supposed to be decidedly liberal.

In the instance just cited no attempt was made to learn how much of the food purchased was actually consumed and how much was rejected. In some of the dietaries published by the Massachusetts bureau such estimates were made. That of a student's club in a New England college will serve as an example.

The young men of the club, some 25 in number, were mostly from the Northern and Eastern States, and, coming from the class of families whose sons go to college, it seems fair to assume that their habits of eating formed at home would not differ materially from those of the more intelligent classes of people in that part of the country. While the diet of the club was substantial and wholesome, it was plain, as was, indeed, necessary, because several of the members were dependent upon their own exertions and the majority had rather limited means. Though fond of athletic sports, they could hardly be credited with as much muscular exercise as the average "laboring man at moderate work." The matron, a very intelligent, capable New England woman, had been selected because of her especial fitness for the care of such an establishment. The steward who purchased the food was a member of the club, and had been chosen as a man of business capacity. He thought that very little of the food was left unconsumed. "All of the meat and other available food that was not actually served to the men at the table," said he, "was carefully saved and made over into croquettes. Men who work their way through college can not afford to throw away their food." But actual examination showed the waste to be considerable. The estimates of the quantities of nutrients were based upon the quantities of food materials for a term of three months and upon the table and kitchen refuse for a week. The results were as follows: In food purchased, protein, 161 grams; energy, 5,345 calories. In waste, protein, 23 grams; energy, 520 calories. In food consumed, protein, 138 grams; energy, 4,825 calories. One-eighth of the protein and one tenth of the energy were simply thrown away.

During the succeeding term a second examination of the dietary of the same club was made. Another steward was then in charge. He had learned of the excessive amounts of food in the former dietary, and planned to reduce the quantities. This was done largely by diminishing the meats. He stated that he did not apprise the club of the change, and that it was not noticed. As he put it, "The boys had all they wanted, and were just as well pleased as if they had had more." Estimates as before, but with more care in determining the waste, showed in food purchased, protein 115 grams; energy, 3,875 calories. In waste, protein, 11 grams; energy, 460 calories. In food consumed, protein, 104 grams; energy, 3,415 calories. One-tenth of the nutritive material of the food this time was thrown away. The young men were amply nourished with three-fifths of the nutrients they had purchased in the previous term.

How much food is required on the average by men whose labor is mainly intellectual is a question to which physiology has not yet given a definite answer, but it is safe to say that the general teaching of the specialists who have given the most attention to the subject would call for little more than the 104 grams of protein and very much less than the 3,400 calories of energy in the food estimated to be actually consumed by these young men when the second examination was made. They could have dispensed with half of all the meats, fish, oysters, eggs, milk, butter, cheese, and sugar purchased for the first dietary and still have had more nutritive material than they consumed in the second. Not only was one-tenth or more of the nutrients thrown away in each of the two cases, but what makes the case still worse peculiarily, the rejected material was very largely from the animal foods in which it is the most expensive.

The estimates of the quantities of food in the two dietaries just quoted were made from tradesmen's bills and the composition was calculated from analyses of similar materials rather than of those actually used. The figures are therefore less reliable than if the food and wastes had been actually weighed and analyzed. In some dietaries lately examined in Middletown, Conn., all the food has been carefully weighed and portions have been analyzed, and the same has been done with the table and kitchen refuse. The results therefore show exactly how much was purchased, consumed, and thrown away. One dietary so investigated was that of a boarding house. The boarders were largely mechanics of superior intelligence and skill, and earning good wages; the mistress was counted an excellent housekeeper and the boarding house a very good one. About one-ninth of the total nutritive ingredients of the food was left in the kitchen and table refuse. The actual waste was worse than this proportion would imply, because it consisted mostly of the protein and fats, which are more costly than the carbohydrates. The waste contained nearly one-fifth of the total protein and fat, and only one-twentieth of the total carbohydrates of the food. Or, to put it in another way, the food purchased contained about 23 per cent more protein, 24 per cent more fats, and 6 per cent more carbohydrates than were eaten. And, worst of all for the pecuniary economy, or lack of economy, the wasted protein and fats were mostly from the meats which supply them in the costliest form.

In another dietary, that of a carpenter's family, also in Middletown, Conn., 7.6 per cent of the total food purchased was left in the kitchen and table wastes. The total waste was somewhat worse than this proportion would imply, because it consisted mostly of the protein and fats, which are more costly than the carbohydrates. The waste contained about one-tenth of the total protein and fat, and only one twenty-fifth of the total carbohydrates of the food; or, to put it in another way, the food purchased contained nearly 10 per cent more protein, 12 per cent more fat, and 5 per cent more carbohydrates than were eaten; and here again the wasted protein and fats were mostly from the meats, which supplied them in the costliest form. At the rate in which the nutrients were actually eaten in this dietary, the protein and fats in the waste would have each supplied one man for a week, and the carbohydrates for three days.

These cases are probably exceptional; at least it is to be hoped that they are. Among 8 dietaries lately studied in Middletown those above named showed the largest proportion of material thrown away. In the rest it was much less. In two cases there was almost none. It is worth noting, however, that the people in these two had the largest incomes of all. In other words, the best-to-do families were the least wasteful.

This form of bad economy is not confined to the kitchen, but begins in the market. In buying meat in the retail markets it is a common practice to have the bone and considerable of the fat cut out and left.

In thus removing the "trimmings" the butcher is apt to cut out considerable else than the bone and fat. In a piece of roast beef weighing 16 pounds, the "trimmings," which consisted of the bone and the meat cut out with it, and which were left for the butcher to sell to the soap man, or get rid of as he might otherwise choose, weighed 4½ pounds. The butcher said that he sold this sort of beef largely to the ordinary people of the city—mechanics, small tradesmen, and laborers.

The 4½ pounds of "trimmings" consisted of, approximately, 2¼ pounds of bone and one-half pound of tendon ("gristle"), which would make a most palatable and nutritious soup, and 1¾ pounds of meat, of which 1 pound was lean and three-fourths pound fat. It is estimated that the nutritive materials of meat thus left unused, saying nothing of the bone and tendon, contained some 15 per cent of the protein and 10 per cent of the potential energy of the whole. The price of the beef was $2.24. Assuming the nutritive value of the ingredients of the "trimmings" to be 12½ per cent of the whole, 28 cents worth of the nutritive material, besides the bone and tendon, was left at the butcher's.

Dr. S. A. Lattimore, professor of chemistry in the University of Rochester, N. Y., tells me that while a member of the board of health of that city he directed the officer in charge of the collection of garbage to note the character of the waste material. It was ascertained that from the streets inhabited by the well-to-do classes, where the culinary affairs were largely left to the servants, the amount of waste thus collected was enormous, and that a considerable proportion of the food purchased was literally thrown away by careless servants. A surprisingly large amount of this waste consisted of good bread. Among the people in moderate circumstances this waste was less.

The common saying that "the average American family wastes as much food as a French family would live upon" is a great exaggeration, but the statistics cited show that there is a great deal of truth in it. Even in some of the most economical families the amount of food wasted, if it could be collected for a month or a year, would prove to be very large, and in many cases the amounts would be little less than enormous.

○